The Responsible AI Redbook

First Edition

The Responsible AI Redbook is a comprehensive guide exploring the ethical, transparent, and accountable deployment of artificial intelligence across industries. It provides actionable frameworks, real-world use cases, and best practices to harness AI's potential responsibly. This essential resource bridges the gap between innovation and ethical integrity, empowering organizations to build trustworthy AI systems that drive positive societal impact.

The Responsible AI Redbook

GenAI was instrumental in shaping the insights and content for 'The Responsible AI Redbook,' providing valuable assistance in researching, structuring, and refining complex concepts into a comprehensive guide for AI enthusiasts and professionals alike.

Book Cover Design: *The cover for this book has been designed by Mugdha Shejale*

I extend my heartfelt gratitude to Gen AI tools for their invaluable assistance in curating the content of The Responsible AI Redbook. Special thanks to Radhika, Anish, and Anshuk for their unwavering support and encouragement throughout this journey. Their contributions have been instrumental in bringing this vision to life.

Shyam Narayan

Contents

Preface

Artificial Intelligence (AI) has transformed from a futuristic concept to a cornerstone of modern society, revolutionizing industries, economies, and human interactions. Yet, with this rapid advancement comes profound ethical, legal, and societal questions. Responsible AI—the practice of developing and deploying AI systems that prioritize fairness, accountability, transparency, and sustainability—is no longer optional. It is an imperative for creating AI systems that align with human values and societal needs.

The Responsible AI Redbook is a definitive guide designed to navigate the complexities of Responsible AI. This book serves as a resource for AI practitioners, policymakers, business leaders, and researchers, offering an in-depth exploration of global regulations, implementation strategies, and best practices. By bridging theory and practice, it equips readers to create and govern AI systems responsibly.

About This Book

This book is meticulously structured to provide readers with a comprehensive understanding of Responsible AI. Each chapter offers critical insights into various facets of Responsible AI, enabling readers to grasp the global context, address implementation challenges, and leverage best practices for success.

- **Chapter 1: Global AI Regulatory Landscape** This chapter sets the stage by mapping the global regulatory landscape. It explores the diverse AI governance frameworks and regulations adopted by countries worldwide, highlighting their similarities, differences, and implications for AI practitioners and organizations.
- **Chapters 2 to 9: Understanding Responsible AI and Regional Regulations** These chapters delve into the foundational principles of Responsible AI and examine regulatory frameworks in key regions, including Europe, the United States, China, India, and more. Readers will gain insights into the unique regulatory challenges and approaches adopted by each region, along with their global impact.

- **Chapter 10: Risks and Impacts** Addressing the potential risks of AI systems, this chapter explores ethical, societal, and operational implications. It discusses risk mitigation strategies to ensure that AI systems are deployed safely and effectively.
- **Chapters 11 to 14: Lifecycle and Design Principles of Responsible AI** These chapters introduce the lifecycle of Responsible AI and provide actionable frameworks for compliance, ethical design principles, and building workflows that integrate ethics and accountability at every stage. Practical methodologies for embedding Responsible AI into organizational processes are detailed.
- **Chapters 15 to 18: Human-In-The-Loop Systems and Supporting Technologies** Practical applications take center stage here, focusing on Human-in-the-Loop (HITL) systems, quality engineering, and the measurement of AI impacts. Readers will also explore cutting-edge tools and technologies that support Responsible AI, including frameworks and platforms from leading cloud providers.
- **Chapters 19 to 23: Best Practices and Governance** These chapters present best practices for implementing Responsible AI across industries and delve into governance models that ensure accountability and oversight. The challenges of operationalizing Responsible AI are candidly discussed, with solutions for overcoming common barriers.
- **Chapters 22 to 26 Future Directions and Real-World Applications** The final chapters inspire readers to look forward, exploring future directions for Responsible AI and providing a call to action. Real-world use cases across industries demonstrate how Responsible AI principles can be applied effectively, showcasing success stories and lessons learned.
- **Glossary** To support understanding, a comprehensive glossary of Responsible AI terms is included, serving as a quick reference for readers.

Who Should Read This Book

This book is designed for a broad audience, including:

- **AI Practitioners and Engineers:** To gain practical guidance on building and deploying Responsible AI systems.
- **Policymakers and Regulators:** To understand the global landscape of AI regulations and their implications.
- **Business Leaders and Executives:** To explore how Responsible AI can drive organizational growth and trust.

- **Researchers and Academics:** To engage with the ethical and technical challenges of AI development.

Whether you are an AI enthusiast or an experienced professional, this book offers valuable insights and actionable frameworks for navigating the evolving world of Responsible AI.

About the Author

Shyam Narayan is a thought leader and innovator in the field of Artificial Intelligence. Shyam has spearheaded complex AI transformation programs across industries. He is passionate about leveraging technology to create value for clients while adhering to ethical and responsible AI principles. Shyam has authored multiple global reports on technology and has been a keynote speaker at various international events. His expertise and commitment to Responsible AI shine throughout this book, making it a must-read for anyone seeking to align AI innovation with ethical practices.

How to Use This Book

This book is designed to be both a reference and a guide. Readers can:

- **Dive Deep:** Explore chapters sequentially for a comprehensive understanding of Responsible AI.
- **Focus on Specific Topics:** Refer to individual chapters for targeted insights, such as regional regulations or best practices.
- **Leverage Frameworks:** Apply the methodologies, tools, and principles outlined in the book to real-world projects.
- **Stay Updated:** Use the glossary and tools sections to keep up with evolving terminology and technologies in Responsible AI.

Acknowledgments

This book is the culmination of extensive research, collaboration, and a shared commitment to advancing Responsible AI. I am deeply grateful to the many individuals and organizations whose contributions made this book possible:

- **Industry Experts and Practitioners:** For sharing their insights and experiences that enriched the content of this book.
- Colleagues & Friends : For their unwavering support and thought leadership in Responsible AI.
- **Policymakers and Regulators:** For their efforts in shaping the global AI regulatory landscape.

- **Family and Friends:** For their encouragement and patience throughout the writing process.

Your collective wisdom and dedication inspire the vision behind this book. Thank you for being part of this journey.

Final Thoughts

As AI continues to shape our world, it is our collective responsibility to ensure its development and deployment align with ethical values. **The Responsible AI Redbook** is a call to action for organizations, leaders, and individuals to prioritize fairness, accountability, transparency, and sustainability in AI systems. By adopting the principles and practices outlined in this book, we can foster trust in AI, mitigate risks, and unlock its full potential for societal good.

Let us embark on this journey together to create a future where AI serves as a force for innovation, equity, and progress—responsibly.

Shyam Narayan

Author, The Responsible AI Redbook

January 2025

Foreward

Foreward By Phil Fersht,

CEO and Chief Analyst, HFS Research

Companies are under increasing pressure from their consumers, partner and regulators to use AI responsibly, making it a strategic necessity rather than an optional add-on. We have to stop using "ethics" as an excuse to avoid investing in AI. Ethical standards are something enterprise leaders must lay out for their enterprises regardless of technology investments.

As AI systems increasingly influence decisions that affect individuals and communities, integrating these ethical considerations is crucial to preventing biases, ensuring privacy, and protecting human rights. Net-net, organizations that prioritize responsible AI practices are will quickly engender trust with their partners, employees, customers and other key stakeholders.

Big kudos to Narayan for addressing these issues at a strategic, pragmatic and technical level.

Foreward By Piyush Sharrma,

CEO & Cofounder - Tuskira.ai a preemptive cyber defense & response platform

Artificial Intelligence (AI) has transformed the cybersecurity industry, becoming both a powerful ally and a formidable adversary. On one hand, AI strengthens cybersecurity by detecting threats faster, automating incident responses, and enabling predictive analysis to thwart potential breaches. On the other, it introduces new challenges, such as adversarial attacks, AI-generated phishing scams, and the weaponization of AI by malicious actors. These dual capabilities make responsible AI practices critical for the cybersecurity industry.

The Responsible AI Redbook provides a comprehensive framework for navigating the ethical, operational, and societal implications of AI in cybersecurity. It equips security professionals, technology leaders, and policymakers with the tools to harness AI's potential while mitigating its risks. From ensuring algorithmic fairness to preventing biases in threat detection models, this book addresses the nuanced challenges of deploying AI responsibly in a domain where accuracy and trust are paramount.

The relevance of responsible AI in cybersecurity cannot be overstated. Cyber defenses powered by AI require transparency and explainability to ensure trust in their decisions, especially when determining what constitutes a threat. Furthermore, as organizations rely more heavily on AI-driven systems, accountability frameworks become essential for managing risks associated with AI errors, data privacy violations, and ethical lapses.

This book delves into actionable strategies for integrating responsible AI practices into cybersecurity workflows. It explores key areas such as bias mitigation, adversarial robustness, ethical threat modeling, and compliance with global security regulations. By providing real-world examples and best practices, it offers a roadmap for building AI solutions that are both innovative and ethically sound.

The Responsible AI Redbook is a timely resource for the cybersecurity industry. It challenges us to not only leverage AI for stronger defenses but to do so responsibly, ensuring that technological advancements align with ethical principles and societal well-being.

Foreward By Ani Filipova,
The Change Advisor for Leaders I Former Bank COO (Citi)

Artificial Intelligence is no longer a concept - it's here; it is a reality that shapes industries, decisions, and lives in a way and pace we could only dream of a decade ago.

I see this firsthand in my daily work. Whether it's ChatGPT for writing and ideation, Perplexity for research, Canva for design, Grammarly for polished communication, or many others, AI tools have become my partners in productivity and creativity.

The corporate training programs I deliver are focused on equipping employees to harness the power of AI effectively - this is what the corporations demand. The question is no longer "if" AI will transform our work, but "how" we can use it responsibly to improve outcomes for people and businesses alike.

These are not just trends but the core of conversations happening in boardrooms, conferences, and workshops that I attend. Leaders are grappling with questions about the ethical use of AI, its long-term societal impact, and the skills their workforce needs to stay ahead in an AI-driven future. The dialogue is no longer limited to "what AI can do" but extends to "what AI should do" and "how we ensure it benefits everyone."

Yet, with every innovation comes an equally monumental responsibility to ensure that AI serves people, has a foundation of ethical principles, and aims to create a better, fairer future for us.

The Responsible AI Redbook is not a technical guide but a call to rethink our approach to innovation. It invites us to contemplate the systems we create and the impact they and we will leave behind. It also reminds us that our choices today will have lasting effects far beyond tomorrow.

In this book, you will find actionable models, real-world cases, and the tools you need to navigate the complex landscape of Responsible AI. Whether you are a decision-maker, a developer, or simply someone passionate about doing the right thing, these pages will inspire you to turn values and principles into practice.

This is not just another book about AI; It is a call to action and a reminder that the choices we make today will ripple through future generations.

If we get this right, we don't just build better technology and systems - we create a better world.

Dive in. Let this be your guide, your challenge, and your commitment to doing AI responsibly.

Foreward By Niranjan Prabhu,

Chief Information Officer | Board Member | Top 50 Most Innovative CIOs 2023/2020

Artificial Intelligence (AI) is revolutionising industries and reshaping society at an unprecedented pace, and its impact on the education sector is nothing short of transformative. From personalised learning platforms to streamlined administrative processes, AI is redefining how we teach, learn, and manage educational systems. Yet, with this immense potential comes an equally significant responsibility: to ensure that AI is developed and deployed ethically, equitably, and with the well-being of all stakeholders in mind. The Responsible AI Redbook emerges as a vital guide, offering a comprehensive framework for navigating the ethical, practical, and societal dimensions of AI in education with the global view. It is an indispensable resource for educators, administrators, policymakers, and technology providers alike.

For the education sector, this book is especially timely and relevant. As the custodians of knowledge and the architects of future innovation, educational institutions and organisations bear a unique responsibility to address pressing questions: How can AI be leveraged to promote equity and inclusion in education? What safeguards are necessary to ensure AI-driven tools are free from bias and accessible to all? And how do we prepare students not only to use AI responsibly but also to become ethical developers and leaders in this rapidly evolving field?

The Responsible AI Redbook provides actionable strategies for integrating responsible AI practices into every facet of education—from curriculum design and teaching methodologies to operational decision-making and policy development. It delves into critical topics such as bias mitigation, algorithmic transparency, data privacy, governance frameworks, and the long-term societal implications of AI. By exploring these themes, the book equips the education industry with the tools needed to harness AI's potential responsibly while fostering trust, transparency, and accountability.

This book demonstrates how ethical AI principles can be embedded into education and other sectors, ensuring that technology serves as a force for good. It challenges the education sector to lead by example, setting a standard for responsible AI adoption that prioritises inclusivity, fairness, and long-term societal benefit. By doing so, we can create learning environments that are not only innovative but also equitable, empowering future generations to navigate the complexities of AI with integrity and purpose.

The Responsible AI Redbook is more than a guide—it is a call to action. It urges the education industry to embrace its role as a pioneer in ethical AI, ensuring that innovation and responsibility go hand in hand. Together, we can build a future where AI enhances education for all, fostering a world that is not only smarter but also more just and compassionate.

Foreward By David Cushman,

Executive Research Leader, HFS Research

As someone deeply immersed in the evolving enterprise AI landscape, I've come to see Responsible AI not as a luxury or a buzzword but as a foundational necessity for any organization aiming to thrive in the AI-driven era. This book captures the urgency and importance of that necessity with precision and practicality.

The rapid rise of AI has unlocked opportunities we couldn't have imagined a decade ago, but it has also surfaced challenges we can no longer afford to ignore—bias, transparency, accountability, and trust, to name just a few. What's clear is that the organizations able to align AI with their core principles and values, while addressing these challenges head-on, will be the ones who win the long game.

This book stands out because it doesn't just dwell on theory or broad platitudes. Instead, it delivers actionable insights and frameworks to operationalize Responsible AI. Whether you're a leader, a technologist, or a strategist, this book equips you to embed ethics and responsibility into the very fabric of your AI initiatives—not as a regulatory checkbox, but as a driver of trust, differentiation, and long-term value.

In my role at HFS Research, I see businesses at the forefront of this journey grappling with the same questions posed here: How do we balance innovation and responsibility? How do we implement ethical guardrails at scale? How do we lead with trust in a time of increasing skepticism? This book offers clarity and answers that will resonate across industries.

If you're serious about using AI to drive sustainable, meaningful impact, this book will inspire you, challenge you, and, most importantly, empower you. It's a privilege to introduce this insightful and essential work to anyone ready to take Responsible AI from concept to competitive advantage.

Chapter 1

Global AI Regulatory Landscape

Introduction

Artificial Intelligence (AI) has transitioned from an emerging technology to a transformative force reshaping industries, economies, and societies worldwide. As its capabilities grow, so do the concerns surrounding its ethical, legal, and societal impacts. Countries around the globe are adopting varied approaches to regulate AI, reflecting their unique priorities, cultural contexts, and governance structures. This chapter explores the evolution of AI across the world, the regulations being implemented in different regions, the differences in approach among countries, and key discussions held at global forums like the World Economic Forum (WEF).

The Evolution of AI Across the World

1. Early Development and Adoption:
   - The evolution of AI began with foundational research in the 1950s, with key milestones such as the development of expert systems and machine learning algorithms.
   - By the 2010s, advances in computing power and data availability led to breakthroughs in deep learning and natural language processing, propelling AI into practical applications across industries like healthcare, finance, and transportation.
2. Regional Pioneers:
   - United States: Home to leading tech companies and research institutions, the U.S. has driven innovation in AI, focusing on commercial applications and technological dominance.

- Europe: Known for its emphasis on ethical AI, Europe has prioritized regulations ensuring privacy, transparency, and accountability.
- Asia: Countries like China and Japan have heavily invested in AI for economic growth, with China leading in AI-powered surveillance and Japan focusing on robotics and automation.

3. Global Trends:
    - Democratization of AI through open-source frameworks.
    - Increasing collaboration between academia, industry, and governments.
    - Rising concerns about bias, explainability, and data security.

Global AI Regulations

As AI adoption accelerates, countries are implementing regulations to balance innovation with ethical considerations. Below are key regulatory efforts:

1. United States:
    - AI Bill of Rights: A voluntary framework outlining principles such as safety, transparency, and protection against algorithmic discrimination.
    - Sector-specific guidelines, such as those by the Federal Trade Commission (FTC), emphasizing fair and transparent AI practices.
2. European Union:
    - EU AI Act: A risk-based framework categorizing AI systems into unacceptable, high, limited, and minimal risks, with stringent requirements for high-risk applications.
    - GDPR: Governs data privacy and imposes limitations on automated decision-making.
3. China:
    - Regulatory focus on ensuring national security and controlling data flows.
    - Ethical guidelines addressing the misuse of AI in sensitive applications like facial recognition and social credit systems.
4. India:
    - Draft frameworks such as Responsible AI for All promote inclusive AI adoption while ensuring data privacy and fairness.
    - Focus on using AI in agriculture, healthcare, and education to drive socio-economic growth.
5. Canada:
    - Algorithmic Impact Assessment (AIA) tool ensures transparency and accountability in government AI systems.

- Policies encourage innovation while addressing risks like bias and privacy violations.

6. Japan:
   - Guidelines emphasizing human-centered AI, promoting harmony between AI technologies and societal values.
   - Investments in robotics and automation as part of a broader AI strategy.
7. Australia:
   - Voluntary AI Ethics Framework encourages organizations to prioritize fairness, accountability, and safety.
   - Regulatory bodies emphasize transparency in AI-powered decision-making.

Differences in Regulatory Approaches

1. Focus Areas:
   - Europe: Prioritizes ethics and human rights, focusing on regulations to protect privacy, prevent bias, and ensure transparency.
   - United States: Emphasizes innovation and competitiveness, with a lighter regulatory touch.
   - China: Uses AI regulation to bolster national security and strengthen state control over data and technology.
   - India: Aims for AI-driven socio-economic development, focusing on accessibility and fairness.
2. Regulatory Models:
   - Prescriptive (Europe): Detailed laws and frameworks to govern specific AI risks and applications.
   - Principle-Based (United States, Australia): Voluntary guidelines providing high-level principles for ethical AI adoption.
   - State-Controlled (China): Strict regulatory oversight aligned with national interests and strategic objectives.
3. Implementation Mechanisms:
   - Europe: Independent oversight bodies to enforce compliance.
   - United States: Industry-led compliance with government guidelines.
   - China: Centralized control and mandatory compliance with state directives.

Key Discussions at the World Economic Forum

The World Economic Forum (WEF) has been instrumental in fostering global dialogue on Responsible AI. Recent discussions have centered on:

1. Harmonizing Global Regulations:
    - Challenges in aligning regional regulations to create a cohesive global framework.
    - The role of international organizations in standardizing AI governance.
2. AI for Sustainable Development:
    - Leveraging AI to address global challenges such as climate change, healthcare, and education.
    - Ensuring equitable access to AI technologies in developing nations.
3. Ethical Considerations:
    - Addressing bias, fairness, and accountability in AI systems.
    - Balancing innovation with societal well-being and individual rights.
4. AI and Workforce Transformation:
    - Preparing for job displacement and upskilling workers for AI-driven economies.
    - Ensuring AI complements human labor rather than replacing it entirely.
5. Public-Private Collaboration:
    - Encouraging partnerships between governments, businesses, and academia to drive Responsible AI adoption.
    - Examples of successful collaborations showcased at WEF events.

Conclusion

The evolution of AI across the world reflects diverse priorities, from ethical considerations in Europe to innovation-driven policies in the United States and state-controlled strategies in China. These varied approaches highlight the complexities of governing a transformative technology. Global forums like the World Economic Forum play a critical role in bridging gaps, fostering dialogue, and creating pathways for harmonized AI governance. For organizations and policymakers, understanding these differences is essential for navigating the regulatory landscape and ensuring Responsible AI development on a global scale.

Chapter 2

Understanding Responsible AI

Introduction

Responsible AI (RAI) has become a cornerstone in discussions about the ethical and effective use of Artificial Intelligence. As defined by OpenAI, Responsible AI is about "developing AI systems that are aligned with human values, ensuring they are used safely and for the benefit of society." According to Wikipedia, Responsible AI refers to "a set of practices and principles aimed at ensuring AI systems are designed and deployed in ways that are ethical, transparent, and accountable." Similarly, the dictionary defines Responsible AI as "the practice of creating AI technologies that are morally sound, unbiased, and compliant with legal and societal norms." These definitions collectively underscore the criticality of aligning AI technologies with ethical, legal, and societal expectations. As AI technologies become integral to various industries, this chapter explores the essence of Responsible AI, the factors driving its prominence, and the implications of having or lacking regulations.

What is Responsible AI?

Responsible AI refers to the practice of creating and deploying AI systems in a manner that is fair, accountable, transparent, and aligned with ethical principles. It encompasses several key dimensions:

Fairness

Ensuring fairness involves addressing the biases inherent in AI systems and their training data. For example, an AI model trained on imbalanced datasets may disproportionately classify certain groups as higher risk. To counteract this:

- Bias Detection: Regularly analyze training datasets for underrepresentation or overrepresentation of specific demographics, regions, or scenarios. This involves employing statistical tools to measure fairness across different groups.
- Bias Mitigation: Implement algorithms that adjust imbalances in training data. For instance, re-sampling underrepresented classes or using fairness-aware learning models can help.
- Inclusive Model Evaluation: Validate AI performance across diverse test cases, ensuring consistent accuracy and fairness for all user groups.

Transparency

Transparency is critical to fostering trust in AI systems. Achieving transparency includes:

- Explainable Outputs: Use techniques like SHAP (SHapley Additive exPlanations) or LIME (Local Interpretable Model-agnostic Explanations) to provide insights into why the AI system made specific decisions.
- Stakeholder Education: Create documentation and tools that explain AI decision-making processes in non-technical terms for business stakeholders, regulators, and end-users.
- Traceability: Maintain logs that document the data sources, preprocessing steps, and decision logic applied during AI operations.

Accountability

Accountability ensures that AI systems operate within defined ethical and legal frameworks. It involves:

- Governance Structures: Establish committees or boards to oversee AI system design, deployment, and performance.
- Error Management Protocols: Develop protocols for addressing system failures. This includes defining escalation paths, remediation plans, and user notifications.
- Regulatory Compliance: Align AI development and usage with standards such as ISO/IEC 38507 or NIST AI Risk Management Framework.

Privacy

Privacy is a cornerstone of Responsible AI, ensuring the safeguarding of user data. Key practices include:

- Data Minimization: Collect only the data necessary for AI tasks, avoiding excessive or irrelevant information.
- Encryption: Use strong encryption standards for both data in transit and at rest to protect against breaches.
- Anonymization: Remove personally identifiable information (PII) before using data for training or inference.
- Compliance: Align practices with laws like GDPR and HIPAA, conducting regular audits to ensure adherence.

Sustainability

Sustainability focuses on minimizing the environmental impact of AI systems. This involves:

- Energy Efficiency: Optimize models and training pipelines to reduce computational overhead. Techniques such as model pruning and quantization can help.
- Renewable Resources: Shift data center operations to renewable energy sources, aligning with green energy initiatives.
- Lifecycle Management: Regularly update and refine AI systems to maintain efficiency, reducing resource waste.

Responsible AI is not merely a technical challenge but also an ethical and societal imperative, requiring collaboration across disciplines and stakeholders.

Why is There So Much Discussion on Responsible AI?

The heightened discourse on Responsible AI arises from the increasing reliance on AI technologies and their profound impact on society. Key factors contributing to this discussion include:

1. High-Profile Failures
    - Numerous cases have highlighted the potential harms of unregulated or poorly governed AI systems:
        - AI hiring tools discriminating against women.
        - Predictive policing systems disproportionately targeting minority communities.
    - Such failures have sparked public outcry and demands for better oversight.

2. Ethical Dilemmas
    - AI is being used in sensitive areas such as healthcare, criminal justice, and education, where ethical considerations are paramount:
        - How should AI balance accuracy with fairness?
        - Who is accountable for harm caused by AI decisions?
    - These dilemmas necessitate ongoing dialogue among technologists, ethicists, policymakers, and society at large.
3. Regulatory Scrutiny
    - Governments and regulatory bodies worldwide are examining AI practices:
        - The EU's AI Act proposes strict rules for high-risk AI applications.
        - Countries like Canada and the U.S. are introducing AI-specific legislation.
    - These efforts reflect the growing recognition of Responsible AI as a policy priority.
4. Public Awareness
    - The integration of AI into daily life has heightened public expectations for transparency and accountability:
        - Consumers demand ethical practices from companies deploying AI systems.
        - Activist groups and NGOs are advocating for Responsible AI on behalf of affected communities.

Why is Responsible AI Important Now?

The urgency of addressing Responsible AI stems from the rapid adoption and scaling of AI technologies across industries and geographies. Key drivers include:

1. Widespread Deployment
    - AI systems are increasingly used in high-stakes environments:
        - Healthcare diagnostics and treatment planning.
        - Autonomous vehicles and public transportation.
        - Financial services, including credit scoring and fraud detection.
    - Errors or biases in these systems can have significant consequences for individuals and societies.
2. Complexity of AI Systems
    - Modern AI models, particularly large language models, function as black boxes:

   - Their decision-making processes are difficult to interpret.
   - Ensuring fairness and accountability requires sophisticated tools and practices.
3. Exponential Growth
   - The speed at which AI technologies are evolving makes it essential to address ethical and legal concerns proactively:
     - Delaying action risks amplifying harms and eroding public trust.
4. Global Competition
   - Countries and organizations are competing to lead in AI innovation:
     - Responsible AI practices ensure that this competition aligns with societal values and ethical norms.

Why is There Confusion About Responsible AI?

Despite its growing importance, Responsible AI remains a source of confusion for several reasons:

1. Lack of Standardization
   - Different organizations and countries define Responsible AI differently:
     - Some emphasize fairness and ethics, while others prioritize compliance and risk management.
   - The absence of universal standards complicates implementation and measurement.
2. Technical Complexity
   - AI systems are inherently complex, involving:
     - Advanced algorithms and vast datasets.
     - Interdependencies across multiple systems and stakeholders.
   - Non-experts often struggle to assess and address these complexities.
3. Competing Interests
   - Balancing innovation with regulation creates tension among stakeholders:
     - Developers seek flexibility to innovate.
     - Regulators prioritize safeguarding societal interests.
4. Evolving Nature of AI
   - AI technologies and applications are continuously evolving:
     - Best practices and guidelines must adapt to new developments.
     - This fluidity contributes to uncertainty.

Developments in Technology and Platforms Necessitating Responsible AI Regulations

The rapid advancements in AI technologies and platforms underscore the critical need for Responsible AI regulations:

1. Generative AI
   - Tools like GPT and DALL-E are revolutionizing content creation but pose risks such as:
     - Misinformation and deepfakes.
     - Copyright infringement and ethical misuse.
2. Autonomous Systems
   - Self-driving cars and drones demand stringent safety and accountability standards:
     - Accidents and misuse highlight the need for robust oversight.
3. Data-Driven Decision Making
   - AI systems influence critical decisions in areas such as hiring and lending:
     - Ensuring fairness and avoiding discrimination are paramount.
4. Scalability and Ubiquity
   - Cloud platforms and APIs enable rapid deployment of AI systems:
     - Without regulations, the risk of unregulated and unethical AI proliferation increases.

What Will Happen If We Don't Have Responsible AI Regulations?

Neglecting Responsible AI regulations could lead to significant risks:

1. Unmitigated Risks
   - Increased occurrences of bias, discrimination, and privacy violations.
   - Erosion of public trust in AI technologies.
2. Legal and Financial Consequences
   - Organizations may face lawsuits and fines for unethical AI practices.
   - Regulatory backlash could stifle innovation and progress.
3. Societal Harm
   - Marginalized communities may bear the brunt of biased AI systems.
   - Broader societal trust in technology could decline.
4. Stifled Innovation
   - Without clear guidelines, organizations may hesitate to innovate for fear of ethical or legal missteps.

What Will Happen If We Have Responsible AI Regulations?

Conversely, implementing Responsible AI regulations offers numerous benefits:

1. Trust and Adoption
    - Clear guidelines foster public trust, encouraging broader adoption of AI technologies.
2. Global Leadership
    - Countries with robust Responsible AI frameworks can lead ethically in the global AI landscape.
3. Ethical Innovation
    - Regulations ensure that AI technologies align with societal values while driving progress.
4. Risk Mitigation
    - Proactive measures reduce legal, ethical, and operational risks, creating a sustainable AI ecosystem.

Conclusion

Responsible AI is no longer a choice but a necessity. As AI technologies continue to evolve, ensuring fairness, transparency, and accountability has become essential. Addressing the challenges and opportunities of Responsible AI requires collaboration, innovation, and regulation. While the journey may be complex, the risks of inaction far outweigh the efforts required. By embracing Responsible AI practices and regulations, we can foster a future where AI technologies drive innovation while upholding equity and ethics.

Chapter 3

European Regulations on Responsible AI

Introduction

Europe has emerged as a global leader in the governance and regulation of Artificial Intelligence (AI), particularly in promoting Responsible AI practices. Through comprehensive legislative frameworks and policy guidelines, the European Union (EU) aims to ensure AI systems are ethical, transparent, and aligned with societal values. This chapter explores the key regulations rolled out by Europe on Responsible AI, their core principles, and their implications for organizations implementing AI. Each section will delve deeper into the context, mechanisms, and real-world examples to provide a thorough understanding of the evolving landscape.

The EU AI Act: A Landmark Regulation

The EU Artificial Intelligence Act (AI Act), proposed in April 2021, represents one of the most significant legislative efforts to regulate AI comprehensively. It establishes a clear risk-based framework, defining obligations for AI developers, deployers, and users. Let's explore its components in greater detail.

1. Risk Categories:
   - Unacceptable Risk: The Act explicitly bans AI applications that pose a threat to safety or fundamental rights. Examples include AI systems for:
     - Social scoring by governments that undermine human dignity.
     - Exploitative manipulation of vulnerable individuals, such as children or those with disabilities.

    - Real-time biometric identification in public spaces, except under narrowly defined legal circumstances like preventing serious crime.
  - High Risk: AI systems operating in critical sectors such as healthcare, transportation, and education fall into this category. They must comply with strict requirements such as:
    - Rigorous data governance to mitigate biases.
    - Documentation that ensures auditability.
    - Human oversight mechanisms to prevent misuse.
  - Limited Risk: These systems, including chatbots and virtual assistants, require minimal interventions. Organizations must disclose when users are interacting with AI to ensure transparency.
  - Minimal Risk: Covers most AI systems like productivity tools and entertainment applications. These are largely exempt from compliance obligations to encourage innovation.
2. Key Provisions:
  - Mandatory Risk Assessments: High-risk systems require extensive testing and documentation before deployment.
  - Data Governance: Ensures datasets used for training AI are unbiased, high-quality, and representative.
  - Transparency Requirements: Users must be informed about AI's involvement in decision-making processes, especially in sensitive contexts.
  - CE Marking for Compliance: Similar to product safety certifications, this demonstrates adherence to regulatory standards.
3. Impact on Organizations:
  - Organizations must create cross-functional teams involving technical, legal, and ethical expertise to meet these requirements.
  - Investments in compliance infrastructure, such as AI monitoring tools and data validation frameworks, become critical.
  - Consumer trust improves as organizations adopt Responsible AI practices, offering competitive advantages in regulated markets.

General Data Protection Regulation (GDPR) and AI

While GDPR primarily focuses on data privacy, its provisions have substantial implications for AI systems. The regulation ensures that AI practices align with fundamental rights and data protection principles.

1. Relevant Provisions:
    - Article 22: Limits automated decision-making that significantly impacts individuals unless explicit consent is provided or the processing is essential for contract execution. For example:
        - Loan applications powered by AI must incorporate human oversight to meet GDPR requirements.
    - Transparency and Consent: Organizations must clearly explain how personal data is collected, stored, and used in AI applications. Users have the right to opt-out of automated processing.
    - Data Minimization and Accuracy: AI models must process only the data necessary for their purpose, and organizations must take measures to correct inaccurate or outdated data.
2. Impact on Organizations:
    - Privacy-by-design becomes a non-negotiable standard for AI development, ensuring compliance is integrated at every stage.
    - Implementation of anonymization and encryption techniques to secure personal data.
    - Global organizations operating in Europe must ensure their AI systems comply with GDPR to avoid severe penalties, fostering greater accountability and trust.

Ethics Guidelines for Trustworthy AI

The EU's High-Level Expert Group on AI developed the Ethics Guidelines for Trustworthy AI to encourage organizations to go beyond legal compliance and prioritize societal welfare.

1. Core Principles:
    - Human Agency and Oversight: AI systems should empower human decision-making rather than undermine it. For instance:
        - In healthcare, AI diagnostic tools must provide insights that enhance a doctor's decision-making capabilities.
    - Fairness and Non-Discrimination: Organizations must actively address and mitigate biases in AI systems, particularly in critical sectors like hiring, lending, and law enforcement.

- Accountability: Clear documentation of AI system designs, decisions, and outcomes ensures traceability and assigns responsibility for potential harms.

2. Impact on Organizations:
   - Incorporating ethical AI frameworks enhances organizational reputation and stakeholder trust.
   - Encourages long-term investment in fairness and transparency, aligning business goals with societal expectations.
   - Builds resilience against future regulatory changes, positioning organizations as leaders in ethical AI deployment.

AI Liability Directive

Complementing the AI Act, the AI Liability Directive provides mechanisms for individuals to seek compensation for damages caused by AI systems.

1. Key Features:
   - Reversal of Burden of Proof: In cases where AI systems cause harm, the provider must demonstrate compliance with regulatory standards.
   - Comprehensive Liability: Extends accountability to developers, deployers, and even end-users in certain cases, ensuring shared responsibility across the AI lifecycle.
2. Impact on Organizations:
   - Necessitates robust testing and validation procedures to mitigate risks.
   - Encourages investments in liability insurance and legal expertise to manage potential claims.
   - Transparency in AI design and deployment becomes paramount, reducing the likelihood of disputes and fostering trust.

Implications for Organizations

The European regulatory landscape presents both significant challenges and opportunities for organizations:

1. Compliance Challenges:
   - Adapting to new regulations requires extensive resource allocation, from hiring compliance officers to implementing monitoring systems.
   - Smaller organizations may face disproportionate costs, potentially necessitating partnerships or external support.

2. Operational Impacts:
   - Redesigning AI workflows to align with Responsible AI principles ensures long-term sustainability.
   - Organizations must establish feedback loops to continuously monitor and improve AI systems.
3. Strategic Opportunities:
   - Adopting Responsible AI not only meets compliance but also builds consumer trust, opening new market opportunities.
   - Ethical AI practices enhance brand reputation and help attract socially conscious customers and partners.

Case Studies: Organizations Responding to EU Regulations

1. Healthcare: A leading European hospital group redesigned its AI-powered diagnostic systems to include explainability features, ensuring compliance with the AI Act while improving patient trust.
2. Finance: A multinational bank implemented bias detection algorithms and established human oversight for its AI-driven loan approval processes, aligning with GDPR and Trustworthy AI principles.
3. Retail: An e-commerce company introduced clear disclosures about AI-driven recommendations, enhancing transparency and aligning with the EU's ethical guidelines.

Checklist of Regulations and Mitigation Strategies

Regulation	Requirement	Mitigation/Compliance Strategy
EU AI Act: High-Risk Systems	Conduct risk assessments and ensure transparency.	Implement rigorous testing, data governance, and explainability tools.
EU AI Act: Limited Risk Systems	Disclose AI usage to users.	Add clear notifications and ensure accessible user education about AI involvement.
EU AI Act: Unacceptable Risk	Prohibition of applications like social scoring or manipulative AI.	Avoid developing or deploying AI systems falling under prohibited use cases.

GDPR: Article 22	Ensure human oversight in automated decision-making processes.	Establish hybrid workflows combining AI with human reviewers for critical decisions.
GDPR: Transparency and Consent	Inform users about data collection and AI processes; secure explicit consent.	Develop user-friendly privacy policies and consent mechanisms; integrate robust data logs.
GDPR: Data Minimization	Limit data usage to only what is necessary and ensure accuracy.	Implement strict data filtering and real-time validation tools for AI training data.
Ethics Guidelines for Trustworthy AI	Promote human agency, fairness, and accountability.	Embed ethical frameworks in AI development lifecycle; monitor bias mitigation efforts and accountability tracking.
AI Liability Directive	Prove compliance when AI systems cause harm; liability extends to developers and deployers.	Maintain thorough documentation, conduct pre- and post-deployment testing, and acquire liability insurance.
Transparency Obligations	Inform users about AI's role in decisions, particularly in high-stakes applications like healthcare.	Design AI outputs with interpretability in mind; provide easy-to-understand explanations of AI's functionality and outcomes.
CE Marking for Compliance	Certify high-risk AI systems meet EU safety and performance standards.	Conduct third-party audits and obtain necessary certifications before deployment.

Conclusion

Europe's leadership in regulating AI underscores the global importance of aligning innovation with societal values. While these regulations pose challenges, they also provide a blueprint for ethical and sustainable AI adoption. By embracing these standards, organizations can not only mitigate risks but also lead the way in establishing trust and accountability in the global AI ecosystem.

Chapter 4

American Regulations on Responsible AI

Introduction

The United States has long been at the forefront of technological innovation, and artificial intelligence (AI) is no exception. However, with the rapid adoption of AI technologies comes a growing recognition of the need for governance to address ethical, legal, and societal concerns. While the U.S. has taken a decentralized approach to AI regulation, focusing on guidelines and sector-specific rules rather than comprehensive federal legislation, the impact on organizations is profound. This chapter explores the key American regulations on Responsible AI, their implications for organizations, and actionable steps for compliance.

The Regulatory Landscape in the United States

The U.S. regulatory approach to AI emphasizes innovation and flexibility while addressing ethical concerns through a combination of guidelines, executive orders, and sector-specific regulations.

1. The AI Bill of Rights (2022):
    - A framework developed by the White House Office of Science and Technology Policy (OSTP), outlining five principles for the responsible design and use of AI systems:
        - **Safe and Effective Systems**: Ensuring AI systems are rigorously tested and monitored for safety and effectiveness.
        - **Algorithmic Discrimination Protections**: Preventing biases that lead to unfair treatment or disparate impacts.

    - **Data Privacy**: Providing individuals with control over their data.
    - **Notice and Explanation**: Ensuring transparency about when and how AI is used.
    - **Human Alternatives, Consideration, and Fallback**: Allowing users to opt-out of AI systems and seek human intervention.

2. Federal Trade Commission (FTC) Guidelines:
   - The FTC emphasizes fairness, transparency, and accountability in AI practices under its authority to enforce consumer protection laws. Key areas include:
     - Preventing deceptive AI claims.
     - Ensuring transparency in data collection and use.
     - Avoiding discriminatory practices.
3. Sector-Specific Regulations:
   - **Healthcare**: The Food and Drug Administration (FDA) regulates AI in medical devices, requiring rigorous validation and monitoring.
   - **Finance**: The Equal Credit Opportunity Act (ECOA) prohibits AI-driven discrimination in credit decisions.
   - **Employment**: The Equal Employment Opportunity Commission (EEOC) oversees the use of AI in hiring and workplace decision-making to prevent bias.
4. State-Level Initiatives:
   - States like California and New York have introduced AI-related privacy and bias regulations, complementing federal efforts.
     - **California Consumer Privacy Act (CCPA)**: Grants individuals rights over their data, impacting AI systems relying on consumer information.
     - **New York City's AI Hiring Law**: Mandates bias audits for AI-driven hiring tools.
5. National Institute of Standards and Technology (NIST) AI Risk Management Framework:
   - Provides a voluntary framework for managing AI risks, focusing on:
     - Validating AI performance.
     - Ensuring explainability and interpretability.
     - Mitigating risks like bias and security vulnerabilities.

Impact on Organizations

The U.S. regulatory landscape requires organizations to navigate a complex web of guidelines and sector-specific rules, with significant implications for their operations:

1. Compliance Costs:
    - Organizations must invest in tools, training, and personnel to meet compliance requirements.
    - Conducting audits and risk assessments can be resource-intensive, particularly for small and medium-sized enterprises (SMEs).
2. Operational Changes:
    - AI systems must be designed and deployed with fairness, transparency, and accountability in mind.
    - Data governance practices need to be strengthened to ensure privacy and mitigate biases.

3. Reputational Risk:
    - Non-compliance or ethical lapses can lead to public backlash, regulatory scrutiny, and legal challenges, affecting brand reputation.
4. Opportunities for Leadership:
    - Proactively adopting Responsible AI practices can enhance consumer trust and position organizations as leaders in ethical innovation.

Checklist of Requirements and Compliance Steps

Requirement	Description	Steps to Comply
Safe and Effective Systems	Ensure AI systems are thoroughly tested for safety and effectiveness.	Conduct rigorous testing and continuous monitoring; document validation processes and outcomes.
Algorithmic Discrimination Protections	Prevent biases and ensure fairness in AI-driven decision-making.	Implement bias detection tools; train AI on diverse, representative datasets; conduct regular audits for fairness.
Data Privacy	Give users control over their personal data and ensure data security.	Establish robust data governance policies; encrypt data; provide clear privacy

		notices and mechanisms for user consent.
Transparency and Explanation	Inform users about AI usage and provide clear explanations of AI decisions.	Develop user-friendly disclosure statements; incorporate explainability features into AI models.
Human Alternatives	Allow users to opt-out of AI systems and access human decision-makers.	Design workflows that integrate human oversight; provide alternative channels for users to address concerns.
Sector-Specific Compliance	Meet industry-specific regulatory requirements (e.g., healthcare, finance, employment).	Work with regulatory experts to ensure compliance; maintain documentation for audits; implement industry-specific safeguards.
Bias Audits for Hiring Tools	Conduct independent audits for AI-driven hiring systems to detect and address biases.	Engage third-party auditors; document findings and corrective actions; update hiring algorithms based on audit results.
NIST Risk Management Framework	Manage risks related to AI deployment and operations.	Adopt NIST guidelines; assess risks at every stage of the AI lifecycle; develop mitigation strategies for identified risks.
FTC Guidelines on Fairness	Avoid deceptive practices and ensure fairness in AI usage.	Regularly review marketing claims; establish clear processes for handling consumer complaints; ensure transparency in data usage.

Steps Organizations Should Take

1. Conduct Comprehensive Risk Assessments:
    - Identify potential risks in AI systems, including biases, privacy concerns, and security vulnerabilities.
    - Use tools and frameworks like the NIST AI Risk Management Framework to structure risk assessments.

2. Establish AI Governance Frameworks:
    - Create internal policies that align with regulatory requirements.
    - Form cross-functional teams comprising legal, technical, and ethical experts to oversee AI projects.
3. Invest in Training and Awareness:
    - Train employees on Responsible AI principles and regulatory requirements.
    - Foster a culture of accountability and ethical innovation.
4. Leverage Third-Party Expertise:
    - Partner with compliance consultants, auditors, and legal experts to navigate complex regulations.
    - Conduct regular external audits to validate compliance efforts.
5. Adopt Transparency and Explainability Practices:
    - Incorporate features that allow users to understand and challenge AI decisions.
    - Develop clear communication strategies to explain AI usage to stakeholders.
6. Monitor and Iterate:
    - Continuously monitor AI systems for compliance and effectiveness.
    - Update policies, processes, and systems in response to evolving regulations and societal expectations.

Conclusion

The decentralized approach to AI regulation in the United States offers organizations flexibility but also requires careful navigation of multiple frameworks and guidelines. By adopting proactive measures to ensure compliance with principles like fairness, transparency, and accountability, organizations can not only mitigate risks but also gain a competitive edge. Responsible AI is not just a regulatory requirement but an opportunity to build trust, foster innovation, and lead in an increasingly AI-driven world.

Chapter 5

Australian Regulations on Responsible AI

Introduction

Australia has positioned itself as a leader in promoting ethical and responsible AI adoption. While its regulatory framework is not as prescriptive as in regions like the European Union, Australia has adopted a principles-based approach to guide organizations in the ethical use of AI. This chapter explores the key regulations and guidelines shaping Responsible AI practices in Australia, their implications for organizations, and actionable steps enterprises can take to ensure compliance.

Australian Regulations and Guidelines for Responsible AI

Australia's approach to AI governance is built on collaboration between government, industry, and academia. Key initiatives and regulations include:

1. Australia's AI Ethics Principles:
   - Released by the Australian Government, these principles provide a high-level framework to guide organizations in the ethical design, development, and use of AI.
   - The eight principles are:
     - Human, Social, and Environmental Well-being: AI should benefit people and the planet.
     - Human-Centered Values: AI should respect human rights, diversity, and dignity.
     - Fairness: AI systems should be inclusive and free from bias.

    - Privacy Protection and Security: AI should safeguard privacy and data.
    - Reliability and Safety: AI should function as intended and minimize risks.
    - Transparency and Explainability: Users should understand how AI systems work.
    - Contestability: People should be able to challenge AI decisions.
    - Accountability: Organizations must be responsible for AI outcomes.
2. Privacy Regulations:
    - The **Privacy Act 1988** governs how organizations collect, use, and manage personal data.
    - AI systems must comply with obligations around transparency, consent, and data protection.
3. State-Level Legislation:
    - States like New South Wales (NSW) have introduced guidelines for AI use in public services, emphasizing fairness, transparency, and accountability.
4. The National AI Action Plan:
    - Focuses on fostering innovation while ensuring Responsible AI practices.
    - Prioritizes workforce readiness, public trust, and ethical governance.
5. Regulatory Sandbox Initiatives:
    - Australia has established regulatory sandboxes to test AI applications in areas like finance and healthcare while ensuring compliance with ethical standards.

Impact on Organizations

The adoption of Responsible AI regulations and guidelines in Australia presents both challenges and opportunities for organizations:

1. Compliance Obligations:
    - Organizations must embed ethical considerations into AI development and deployment processes.
    - Privacy protection and data governance are critical to meeting regulatory requirements.
2. Operational Changes:
    - Designing AI systems with transparency, fairness, and accountability demands significant investment in technology and training.

- Robust risk management frameworks are essential to identify and mitigate potential harms.

3. Trust and Reputation:
   - Adopting Responsible AI practices enhances stakeholder trust and strengthens an organization's reputation.
   - Non-compliance or ethical lapses can result in regulatory scrutiny and reputational damage.
4. Innovation and Market Leadership:
   - Proactively aligning with Australia's AI Ethics Principles positions organizations as leaders in ethical innovation.
   - Ethical AI can become a competitive advantage in attracting socially conscious consumers and partners.

Checklist of Regulatory Requirements and Compliance Actions

Requirement	Description	Steps for Compliance
Adherence to AI Ethics Principles	Ensure AI aligns with Australia's eight AI Ethics Principles.	Conduct ethical impact assessments; integrate fairness, safety, and transparency into AI design.
Privacy Protection	Comply with the Privacy Act 1988 for managing personal data.	Implement privacy-by-design practices; secure explicit consent for data use; encrypt sensitive data.
Fairness and Bias Mitigation	Ensure AI systems are inclusive and free from bias.	Use diverse training datasets; regularly audit AI models for bias; document actions to mitigate identified biases.
Transparency and Explainability	Provide clear explanations of AI decisions to users.	Incorporate explainability features in AI systems; develop user-friendly documentation and communication strategies.
Accountability Mechanisms	Establish processes to take responsibility for AI outcomes.	Create internal governance structures; assign accountability to teams for AI decision-making and impacts.

Data Security and Risk Management	Protect data and minimize risks associated with AI systems.	Conduct security audits; implement robust access controls; develop incident response plans.
Contestability	Allow users to challenge AI decisions and seek human intervention.	Design workflows enabling human oversight; provide mechanisms for users to dispute AI-driven decisions.
Ethical Testing in Regulatory Sandboxes	Test AI applications in controlled environments to ensure compliance before full deployment.	Participate in regulatory sandbox programs; collaborate with regulators and industry experts during testing phases.
Workforce Training and Awareness	Train employees on Responsible AI practices and ethical guidelines.	Develop training programs; foster a culture of accountability and ethical awareness.

Steps Organizations Should Take

1. Develop an AI Governance Framework:
   - Establish internal policies and procedures aligned with Australia's AI Ethics Principles.
   - Form an ethics committee or assign an AI governance officer to oversee compliance efforts.
2. Implement Ethical Impact Assessments:
   - Assess the potential social, economic, and environmental impacts of AI systems.
   - Identify risks and develop strategies to mitigate harms.
3. Foster Transparency and Accountability:
   - Ensure users understand how AI systems work and their implications.
   - Maintain detailed records of AI design, deployment, and outcomes to enable accountability.
4. Adopt Privacy-by-Design Practices:
   - Integrate privacy protections into the AI development lifecycle.
   - Regularly review data collection, storage, and usage practices for compliance with the Privacy Act 1988.

5. Engage in Continuous Monitoring and Improvement:
    - Monitor AI systems post-deployment to ensure ongoing compliance and effectiveness.
    - Update policies and practices to reflect evolving regulations and societal expectations.
6. Participate in Regulatory Sandboxes:
    - Use sandboxes to test AI systems in a controlled environment, gaining insights and feedback for improvement.
    - Engage with regulators to ensure compliance before full-scale deployment.

Conclusion

Australia's principles-based approach to Responsible AI provides organizations with flexibility while emphasizing the need for ethical and accountable AI practices. By aligning with the AI Ethics Principles, Privacy Act, and other regulatory guidelines, enterprises can build trust, mitigate risks, and position themselves as leaders in ethical innovation. Responsible AI is not just about compliance; it is about contributing to a fair, transparent, and sustainable future in the AI-driven world.

Chapter 6

China Regulations on Responsible AI

Introduction

China is at the forefront of Artificial Intelligence (AI) development, leveraging its vast data resources, government investment, and technological expertise to become a global AI leader. However, the rapid adoption of AI technologies has raised significant concerns about privacy, security, and ethical implications. In response, the Chinese government has implemented a range of regulations and guidelines to ensure Responsible AI development. This chapter examines China's regulatory framework, its impact on organizations, and the steps enterprises need to take to comply with these regulations.

China's Regulatory Framework for Responsible AI

China's approach to AI regulation emphasizes state control, national security, and the alignment of AI applications with societal and economic goals. Key regulations and guidelines include:

1. The New Generation AI Development Plan (2017):
   - This strategic plan outlines China's vision to become a global AI leader by 2030.
   - Emphasizes the ethical and secure use of AI technologies, particularly in sensitive areas such as facial recognition and autonomous systems.
2. Guidelines for AI Ethics (2021):
   - Issued by the Ministry of Science and Technology, these guidelines emphasize principles such as:
     - **Harmony and Human Well-being**: AI should enhance societal welfare and align with cultural values.

   - **Fairness and Justice**: Prevent discrimination and ensure inclusivity in AI systems.
   - **Privacy and Security**: Protect personal data and ensure the safety of AI systems.
   - **Control and Accountability**: Maintain human oversight over AI systems.

3. Data Security Law (DSL) and Personal Information Protection Law (PIPL):
   - **DSL (2021)**: Focuses on data governance, security, and cross-border data flows.
   - **PIPL (2021)**: China's equivalent to the EU's GDPR, regulating the collection, storage, and processing of personal data.
4. Regulations for Deep Synthesis Technology (2022):
   - Addresses the ethical and security challenges of technologies like deepfakes and generative AI.
   - Requires clear disclosure of AI-generated content and measures to prevent misuse.
5. Algorithmic Accountability Guidelines (2022):
   - Issued by the Cyberspace Administration of China (CAC), these guidelines mandate:
     - Transparency in algorithmic decision-making.
     - User rights to opt-out of algorithm-driven personalization.
     - Regular audits to ensure compliance with ethical and legal standards.

Impact on Organizations

The regulatory landscape in China imposes significant responsibilities on organizations developing and deploying AI systems:

1. Compliance Burdens:
   - Companies must implement stringent data governance practices to comply with PIPL and DSL requirements.
   - Regular audits and documentation are required to demonstrate compliance with algorithmic accountability standards.
2. Operational Adjustments:
   - AI systems must incorporate transparency, fairness, and privacy protections into their design.
   - Organizations need to invest in robust security measures to prevent data breaches and misuse of AI technologies.

3. Reputational Risk:
   - Non-compliance can lead to severe penalties, loss of business licenses, and reputational damage.
   - Adopting Responsible AI practices can enhance trust among consumers and stakeholders.
4. Opportunities for Innovation:
   - Compliance with China's regulations can drive innovation in AI governance and ethical design.
   - Companies that align with these standards are better positioned to compete in China's rapidly growing AI market.

Checklist of Regulatory Requirements and Compliance Actions

Requirement	Description	Steps for Compliance
Data Governance (DSL and PIPL)	Ensure secure handling and storage of personal and sensitive data.	Implement data encryption; establish protocols for cross-border data transfers; conduct regular data audits.
Transparency in Algorithms	Disclose how algorithms make decisions and offer user control options.	Develop user-friendly explanations for algorithmic processes; provide opt-out mechanisms for personalization.
Fairness and Bias Mitigation	Ensure AI systems do not discriminate against individuals or groups.	Use diverse datasets for training; regularly audit algorithms for bias and take corrective actions when needed.
AI-Generated Content Disclosure	Clearly label AI-generated content to avoid deception.	Include watermarks or disclaimers on AI-generated media; educate users about identifying such content.
Ethical AI Guidelines	Align AI systems with principles of harmony, fairness, and accountability.	Conduct ethical impact assessments; integrate ethical considerations into AI design and deployment workflows.

Algorithmic Audits	Perform regular audits to ensure compliance with CAC guidelines.	Engage third-party auditors; document findings and improvements; establish internal audit teams for ongoing review.
Cybersecurity and Risk Management	Protect AI systems from unauthorized access and misuse.	Deploy robust cybersecurity measures; implement continuous monitoring and response mechanisms for threats.
Human Oversight	Maintain human control over critical AI decisions.	Design workflows that integrate human-in-the-loop processes; provide clear escalation paths for disputes or errors.

Steps Organizations Should Take

1. Develop a Comprehensive AI Governance Framework:
    - Align internal policies with China's ethical guidelines and regulatory requirements.
    - Assign dedicated teams or officers to oversee AI governance and compliance efforts.
2. Invest in Data Security and Privacy Protections:
    - Implement advanced encryption, anonymization, and access control measures.
    - Regularly review and update data handling practices to align with DSL and PIPL standards.
3. Enhance Algorithmic Transparency and Accountability:
    - Build explainability into AI systems to ensure users understand their decision-making processes.
    - Document the development and deployment of algorithms for auditing purposes.
4. Conduct Ethical and Risk Assessments:
    - Evaluate the societal and ethical impacts of AI systems before deployment.
    - Identify potential risks and develop mitigation strategies to address them.

5. Participate in Regulatory Engagements:
   - Engage with regulators and industry groups to stay updated on evolving requirements.
   - Use regulatory sandboxes to test AI applications and receive feedback on compliance.
6. Foster a Culture of Responsibility:
   - Train employees on Responsible AI principles and the importance of compliance.
   - Encourage ethical decision-making and accountability at all levels of the organization.

Conclusion

China's regulations on Responsible AI reflect its commitment to balancing innovation with societal well-being and national security. While these regulations present compliance challenges, they also drive organizations to adopt robust governance practices and ethical frameworks. By aligning with China's regulatory standards, enterprises can not only mitigate risks but also position themselves as trusted and innovative players in one of the world's largest AI markets. Responsible AI is not just a regulatory necessity in China; it is a foundation for sustainable and equitable technological progress.

Chapter 7

India Regulations on Responsible AI

Introduction

India, with its rapidly growing digital economy and innovation-driven ecosystem, recognizes the transformative potential of Artificial Intelligence (AI). The government has emphasized the importance of Responsible AI in driving socio-economic growth while addressing ethical concerns and data privacy. While India's regulatory framework for AI is still evolving, several key initiatives and policies shape the nation's approach to AI governance. This chapter explores the regulations and guidelines influencing Responsible AI in India, their implications for organizations, and actionable steps enterprises can take to comply with these regulations.

India's Regulatory Framework for Responsible AI

India's approach to AI governance is built on inclusivity, transparency, and ethical considerations, with a focus on leveraging AI for public good. Key regulations and initiatives include:

1. National Strategy for Artificial Intelligence (NSAI):
    - Released by NITI Aayog, the NSAI outlines India's vision for AI adoption across critical sectors such as healthcare, education, agriculture, and infrastructure.
    - Highlights the need for Responsible AI principles, including fairness, transparency, and accountability.
2. Data Protection Bill (DPB):
    - A draft legislation aimed at regulating personal data processing.
    - Proposes obligations for data fiduciaries to ensure transparency, accountability, and user consent in AI systems.

  - Focuses on principles such as data minimization, storage limitation, and purpose limitation.
3. AI for All Initiative:
   - Advocates for inclusive AI adoption, ensuring accessibility and fairness.
   - Encourages the development of ethical AI systems that promote social welfare and reduce inequality.
4. Guidelines for Ethical AI:
   - Released by NITI Aayog, these guidelines emphasize:
     - **Transparency**: AI systems must disclose their purpose and functioning to users.
     - **Fairness**: Prevent discrimination and bias in AI decision-making.
     - **Accountability**: Organizations must ensure clear responsibility for AI outcomes.
     - **Privacy**: Safeguard personal data and respect user consent.
5. Sector-Specific Policies:
   - **Healthcare**: AI applications must comply with ethical guidelines, ensuring patient privacy and safety.
   - **Finance**: AI-driven credit scoring and lending must adhere to principles of fairness and non-discrimination.

6. Digital Personal Data Protection Act (DPDP Act) (2023):
   - Enforces stricter obligations for processing personal data, requiring AI systems to handle data responsibly and transparently.

Impact on Organizations

The evolving regulatory landscape in India places significant responsibilities on organizations:

1. Compliance Requirements:
   - Enterprises must implement robust data governance frameworks to align with data protection laws.
   - Ethical AI design is no longer optional, requiring fairness, transparency, and accountability in all AI systems.
2. Operational Adjustments:
   - Organizations need to redesign workflows to include privacy and bias mitigation measures.
   - Regular audits and risk assessments are essential to maintain compliance.

3. Reputational Considerations:
    - Adopting Responsible AI practices enhances trust among customers, regulators, and stakeholders.
    - Non-compliance can lead to legal penalties and reputational damage.
4. Opportunities for Innovation:
    - Aligning with ethical AI guidelines opens avenues for innovation in areas like healthcare, education, and agriculture.
    - Ethical AI can become a differentiator in competitive markets.

Checklist of Regulatory Requirements and Compliance Actions

Requirement	Description	Steps for Compliance
Transparency and Explainability	Disclose AI system functioning and decision-making processes.	Incorporate explainability features in AI models; provide clear user documentation and communication.
Data Protection Compliance	Align AI systems with the DPDP Act and other data protection laws.	Implement data encryption; obtain explicit user consent for data usage; regularly review and update data policies.
Fairness and Bias Mitigation	Ensure AI systems do not discriminate or produce biased outcomes.	Train AI models on diverse datasets; conduct regular audits to detect and mitigate bias; document corrective actions.
Accountability Mechanisms	Assign clear responsibility for AI decisions and outcomes.	Develop internal governance structures; assign accountability roles within AI teams; maintain records for audits.
Purpose Limitation	Ensure AI systems use data strictly for the intended and disclosed purposes.	Clearly define and communicate data usage policies; implement automated monitoring for unauthorized data usage.

Human Oversight	Integrate human-in-the-loop processes in critical AI decisions.	Design workflows enabling human intervention; establish escalation paths for disputes or errors.
Sector-Specific Compliance	Adhere to guidelines specific to industries such as healthcare and finance.	Collaborate with regulators for sector-specific compliance; update AI systems to reflect industry standards.
Ethical AI Guidelines	Develop AI systems aligned with NITI Aayog's ethical principles.	Conduct ethical impact assessments; prioritize fairness, privacy, and accountability during AI development.
Risk Management and Security	Safeguard AI systems against misuse, breaches, and failures.	Conduct regular security audits; deploy advanced cybersecurity measures; develop incident response plans.

Steps Organizations Should Take

1. Develop an AI Governance Framework:
    - Establish internal policies that align with India's ethical AI guidelines and data protection laws.
    - Form AI ethics committees to oversee governance efforts and ensure compliance.
2. Invest in Training and Awareness:
    - Train employees on Responsible AI principles and the regulatory requirements.
    - Create a culture of accountability and ethical innovation across all teams.
3. Conduct Regular Risk and Ethical Assessments:
    - Evaluate AI systems for ethical concerns and potential risks before deployment.
    - Develop mitigation strategies to address identified risks.
4. Enhance Transparency and Accountability:
    - Build explainability into AI systems, ensuring users understand their decision-making processes.

- Document all aspects of AI development and deployment for regulatory audits.

5. Collaborate with Regulators and Industry Bodies:
   - Engage with policymakers and industry groups to stay informed about evolving regulations.
   - Participate in public-private partnerships to influence and adapt to regulatory changes.
6. Leverage Technology for Compliance:
   - Use automated tools for data governance, bias detection, and compliance monitoring.
   - Adopt privacy-by-design principles to integrate compliance into AI workflows.

Conclusion

India's focus on Responsible AI reflects its commitment to balancing innovation with societal welfare and ethical governance. While the regulatory environment is still developing, organizations must take proactive steps to align with ethical principles and compliance requirements. By embracing transparency, fairness, and accountability, enterprises can not only mitigate risks but also unlock new opportunities in India's rapidly evolving AI landscape. Responsible AI is not just a regulatory necessity but a foundation for long-term trust, growth, and sustainability.

Chapter 8

Canada Regulations on Responsible AI

Introduction

Canada has been a global advocate for ethical and responsible AI practices. With its commitment to balancing innovation with accountability, Canada has introduced a series of regulations and guidelines aimed at ensuring the responsible development, deployment, and usage of Artificial Intelligence (AI). These regulations not only provide a framework for compliance but also help organizations align their AI strategies with societal values and ethical standards. This chapter explores the key Canadian regulations on Responsible AI, their implications for organizations, and a comprehensive checklist to ensure compliance.

Key Regulations on Responsible AI in Canada

1. The Artificial Intelligence and Data Act (AIDA)
   - **Overview**: AIDA is a federal legislation aimed at regulating high-impact AI systems to promote transparency, accountability, and human rights.
   - **Key Provisions**:
     - Organizations must conduct impact assessments for AI systems.
     - Implementation of risk management frameworks.
     - Mandatory disclosure of AI system usage in high-stakes environments.
     - Establishment of governance mechanisms for algorithmic accountability.
     - Severe penalties for non-compliance, including fines and sanctions.

2. Privacy Regulations
   - **Overview**: The Personal Information Protection and Electronic Documents Act (PIPEDA) governs how organizations collect, use, and disclose personal data.
   - **Key Provisions**:
     - AI systems must comply with data minimization and purpose limitation principles.
     - Explicit consent is required for data collection.
     - Privacy impact assessments (PIAs) are recommended for AI implementations.
     - Organizations must adopt robust data protection practices to ensure compliance.
3. Canadian Charter of Rights and Freedoms
   - **Overview**: The Charter protects individuals from discrimination and ensures fundamental rights are upheld in the deployment of AI systems.
   - **Key Implications**:
     - AI systems must be free from bias and discrimination.
     - Organizations must ensure fairness and equality in AI-driven decision-making.
     - Ethical reviews and audits are essential to demonstrate compliance.
4. Industry-Specific Guidelines
   - **Healthcare**: Health Canada has specific guidelines for AI in medical devices, emphasizing safety and efficacy.
   - **Finance**: The Office of the Superintendent of Financial Institutions (OSFI) mandates AI governance to manage risks in financial systems.
5. Standards and Certifications
   - **ISO/IEC Standards**: Canada encourages adherence to international standards, such as ISO/IEC 27001 for information security and ISO/IEC 38505 for governance of IT.
   - **AI Standards Development**: The Canadian Standards Association (CSA) develops AI standards focusing on ethical AI design and deployment.

Impact on Organizations

1. Operational Changes

   Organizations must adopt robust governance structures to align with AIDA requirements, including appointing AI compliance officers and conducting regular audits.

2. Increased Transparency

   AI systems must be transparent in their operations. This includes documenting decision-making processes and providing explanations for AI-driven outcomes.

3. Risk Mitigation

   Implementing risk management frameworks becomes crucial to identify and mitigate potential harm from AI systems.

4. Ethical Accountability

   Organizations are held accountable for the ethical implications of their AI systems, necessitating ongoing training and awareness programs.

5. Cost Implications

   Compliance may involve additional costs for audits, assessments, and technological upgrades but ensures long-term sustainability and trustworthiness.

6. Competitive Advantage

   Adherence to Canadian AI regulations positions organizations as leaders in ethical AI, enhancing their reputation and appeal to stakeholders.

Checklist for Regulatory Compliance

Regulation Requirement	What Enterprises Should Do
Conduct AI system impact assessments	Perform detailed assessments to identify potential risks and impacts of AI systems.
Develop and implement risk management frameworks	Establish risk management protocols, including monitoring and mitigation strategies.
Ensure data collection complies with PIPEDA	Adopt practices for data minimization, purpose limitation, and explicit consent for data collection.

Regularly audit AI systems for biases and fairness	Implement auditing tools and methodologies to identify and correct biases in AI systems.
Maintain transparency in AI usage	Document decision-making processes and provide accessible explanations for AI-driven outcomes.
Align AI systems with the Canadian Charter	Conduct ethical reviews and implement safeguards to ensure fairness and equality.
Adhere to industry-specific guidelines	Follow regulations specific to industries, such as healthcare or finance, and implement best practices.
Obtain necessary certifications	Align with ISO/IEC standards and pursue certifications to validate responsible AI practices.
Appoint an AI ethics officer	Designate leadership roles for governance and oversight of AI initiatives.
Educate employees on Responsible AI	Conduct training programs to raise awareness and ensure compliance among staff.
Engage stakeholders and communities	Involve users, regulators, and affected communities in the development and review of AI systems.
Use explainable AI (XAI) techniques	Implement tools and methodologies to enhance the interpretability and transparency of AI models.
Conduct privacy impact assessments	Regularly evaluate AI systems to ensure compliance with privacy standards and principles.

Conclusion

Canada's regulations on Responsible AI emphasize the importance of transparency, accountability, and ethical considerations in the development and deployment of AI systems. By adhering to these regulations, organizations not only ensure compliance but also foster trust and innovation in the rapidly evolving AI landscape. The provided checklist offers a pragmatic approach for enterprises to align with Canadian regulatory requirements and build Responsible AI systems that resonate with societal values.

Chapter 9

UK Regulations on Responsible AI

Introduction

The United Kingdom is at the forefront of shaping policies and regulations for Responsible AI to ensure ethical innovation and economic growth. With a focus on transparency, accountability, and fairness, the UK's regulatory framework aims to mitigate risks while fostering innovation. This chapter explores the key UK regulations and guidelines for Responsible AI, their impact on organizations, and a checklist to help enterprises ensure compliance.

Key Regulations on Responsible AI in the UK

1. AI Regulation Policy Paper (2023)
   - **Overview**: The UK government's policy paper outlines a pro-innovation approach to AI regulation, focusing on building public trust.
   - **Key Principles**:
     - Safety and security of AI systems.
     - Transparency and explainability.
     - Fairness in AI decision-making.
     - Accountability and governance.
     - Contestability and redress mechanisms.
2. UK GDPR (General Data Protection Regulation)
   - **Overview**: Similar to the EU GDPR, the UK GDPR provides a legal framework for data protection and privacy.
   - **Key Provisions**:
     - Requires explicit consent for data collection and processing.
     - Limits the use of personal data for AI training.

  - Mandates data minimization and purpose limitation.
  - Provides individuals with the right to explanation for AI decisions.

3. Equality Act 2010
   - **Overview**: This Act ensures that AI systems do not lead to discrimination based on protected characteristics such as age, gender, or race.
   - **Key Implications**:
     - Organizations must audit AI systems for biases.
     - Implement fairness checks in AI models.
4. Digital Regulation Cooperation Forum (DRCF)
   - **Overview**: The DRCF coordinates regulatory efforts across UK sectors, focusing on digital technologies, including AI.
   - **Key Focus Areas**:
     - Collaboration among regulatory bodies like Ofcom, ICO, CMA, and FCA.
     - Harmonized guidelines for AI system audits.
5. AI Standards Hub
   - **Overview**: The UK's AI Standards Hub promotes adherence to international standards for Responsible AI.
   - **Key Standards**:
     - ISO/IEC standards for AI safety and governance.
     - Development of UK-specific ethical guidelines.

6. Algorithmic Transparency Standard
   - **Overview**: This standard promotes transparency in public sector AI systems.
   - **Key Requirements**:
     - Disclosure of algorithms used.
     - Clear documentation of decision-making processes.
     - Regular audits for accountability.

Impact on Organizations

1. Governance and Accountability

   Organizations must establish robust governance frameworks, including appointing AI ethics officers and creating oversight committees.

2. Transparency Requirements

   Businesses need to document AI systems' decision-making processes and ensure they can provide explanations to stakeholders.

3. Risk Management

   Enterprises are required to conduct regular risk assessments and implement measures to mitigate potential harms.

4. Bias and Fairness

   AI models must undergo regular audits to ensure they are free from bias and comply with the Equality Act.

5. Data Privacy Compliance

   Organizations must align with UK GDPR requirements for data collection, storage, and processing.

6. Cost Implications

   Compliance may involve investment in audits, training, and technology upgrades but enhances trust and long-term competitiveness.

Checklist for Regulatory Compliance

Regulation Requirement	**What Enterprises Should Do**
Adhere to AI regulatory principles	Implement governance frameworks to align with safety, transparency, and fairness principles.
Conduct AI impact assessments	Evaluate potential risks and societal impacts of AI systems before deployment.
Ensure compliance with UK GDPR	Obtain explicit consent for data usage, minimize data collection, and ensure purpose limitation.
Perform bias audits	Use fairness metrics and tools to detect and mitigate biases in AI models.
Maintain algorithmic transparency	Document and disclose decision-making processes for AI systems.
Establish accountability mechanisms	Appoint AI ethics officers and create governance structures for oversight.

Align with Equality Act 2010	Audit AI systems to prevent discrimination and promote inclusivity.
Follow AI Standards Hub guidelines	Adhere to international and UK-specific AI standards for ethical AI implementation.
Monitor and audit AI systems	Regularly audit AI models for fairness, accuracy, and compliance with regulatory standards.
Educate employees on AI regulations	Conduct training sessions to raise awareness about Responsible AI practices and legal requirements.
Engage stakeholders	Involve users, regulators, and affected communities in evaluating AI system impacts.
Prepare contestability and redress mechanisms	Establish processes to address challenges and complaints related to AI-driven decisions.
Utilize explainable AI (XAI)	Deploy methods to make AI decision-making processes interpretable and understandable.

Conclusion

The UK's regulatory framework for Responsible AI is a balanced approach to promoting innovation while ensuring accountability and fairness. By adhering to these regulations, organizations can mitigate risks, build trust, and leverage AI responsibly. The provided checklist serves as a practical guide for enterprises to align their AI initiatives with UK regulations and drive ethical AI adoption.

Chapter 10

Risks and Impacts

Introduction

The failure to adopt Responsible AI (RAI) practices poses significant risks to organizations, industries, and societies. These risks are not only financial but also ethical, reputational, and operational, threatening the very foundation of trust and innovation in AI systems. Across industries and geographies, the repercussions of neglecting Responsible AI practices have manifested in high-profile failures, regulatory fines, and social backlash. This chapter explores these risks and their impacts, supported by examples from various sectors and regions, emphasizing the critical need for Responsible AI practices.

Key Risks of Not Adopting Responsible AI Practices

1. Bias and Discrimination

   AI systems can inadvertently perpetuate or amplify biases present in training data, leading to unfair outcomes.

   - Impact:
     - **Loss of trust among customers and stakeholders**: Unfair outcomes undermine confidence in AI systems, discouraging adoption.
     - **Legal actions under anti-discrimination laws**: Organizations face lawsuits and fines for deploying biased systems.
     - **Damaged reputation and reduced market competitiveness**: Public backlash against perceived unethical practices can diminish brand equity.

- Example:
  - **Geography**: United States.
  - **Industry**: Hiring platforms.
  - **Case**: An AI-based recruitment tool developed by a tech giant was found to favor male candidates over female candidates due to biased training data. The tool was eventually scrapped, but the incident highlighted systemic issues in training and deploying biased AI systems. This event triggered public scrutiny and led to tighter regulations on AI use in hiring.

2. Privacy Violations

   AI systems that lack robust privacy protections can misuse sensitive data, resulting in violations of privacy laws and loss of consumer trust.

   - Impact:
     - **Regulatory fines and penalties**: Violations of data protection laws like GDPR result in significant financial losses.
     - **Legal liabilities and lawsuits**: Customers or entities affected by privacy breaches may sue for damages.
     - **Loss of customer base and business opportunities**: Customers withdraw trust and loyalty from non-compliant organizations.
   - Example:
     - **Geography**: Europe.
     - **Industry**: Retail.
     - **Case**: A leading retailer's AI-driven personalization engine collected and used customer data without proper consent, violating GDPR. The company faced a substantial fine and customer backlash, leading to a decline in revenue and brand perception.

3. Operational Risks

   Deploying AI systems without Responsible AI practices can lead to operational inefficiencies, system failures, and financial losses.

   - Impact:
     - **Disrupted business operations**: Faulty AI outputs can interrupt critical processes.
     - **Increased maintenance costs**: Errors in AI systems necessitate costly corrections and updates.
     - **Revenue loss due to unreliable systems**: Untrustworthy AI systems drive away users and clients.

- Example:
  - Geography: Asia.
  - Industry: Finance.
  - Case: A bank's credit scoring AI erroneously rejected loan applications due to data quality issues, leading to a drop in customer satisfaction and regulatory scrutiny. Correcting these errors required extensive retraining and system redesign.

4. Regulatory and Legal Risks
   Neglecting RAI can lead to violations of regulatory standards, resulting in fines, sanctions, and legal proceedings.
   - Impact:
     - **Financial penalties**: Non-compliance with laws like AIDA or GDPR can result in multi-million-dollar fines.
     - **Increased regulatory oversight**: Organizations found in violation may face ongoing scrutiny, delaying operations.
     - **Loss of licenses or operational restrictions**: Regulatory bodies can impose bans or revoke licenses.
   - Example:
     - **Geography**: Canada.
     - **Industry**: Healthcare.
     - **Case**: An AI-powered diagnostic tool was found to lack transparency in decision-making, raising concerns under Canada's health data regulations. The tool's use was suspended, impacting service delivery and patient care.
5. Reputational Damage
   Unethical or non-compliant AI practices can severely tarnish a company's reputation.
   - Impact:
     - **Loss of brand equity**: Negative publicity erodes customer trust.
     - **Decline in investor confidence**: Investors shy away from companies embroiled in controversy.
     - **Challenges in attracting talent**: Professionals may avoid working for organizations perceived as unethical.
   - Example:
     - **Geography**: Australia.
     - **Industry**: Insurance.

    - **Case**: An AI-driven claims processing system unfairly rejected valid claims, resulting in public outrage and media scrutiny. The insurer's reputation suffered, and they faced customer attrition and regulatory intervention.

6. Ethical Risks

   AI systems that fail to adhere to ethical guidelines can undermine societal trust and amplify inequalities.

   - Impact:
     - **Public backlash and protests**: Unethical AI systems invite societal rejection and resistance.
     - **Decreased adoption of AI technologies**: Skepticism about AI's role in society can stifle innovation.
     - **Increased societal inequalities and harm**: Marginalized communities may bear the brunt of poorly designed AI systems.
   - Example:
     - **Geography**: South Africa.
     - **Industry**: Education.
     - **Case**: An AI-based education platform disproportionately disadvantaged students from underrepresented communities by tailoring resources inequitably. The incident sparked a national debate on algorithmic fairness and equity in education.

Global Examples of Risks and Impacts

North America

- **Sector**: Policing.
- **Case**: Predictive policing tools in the U.S. were found to disproportionately target minority communities, perpetuating systemic biases and eroding public trust in law enforcement. Activists and researchers called for greater oversight and accountability in law enforcement AI systems.

Europe

- **Sector**: Automotive.
- **Case**: An autonomous vehicle in Germany caused a fatal accident due to insufficient safety measures. This incident led to stricter regulations on self-driving technology and public skepticism about its safety.

Asia

- **Sector**: Social Media.
- **Case**: A social media platform in India faced backlash for spreading misinformation through its recommendation algorithms. The resulting legal action and increased regulatory oversight forced the platform to revise its algorithms.

Africa

- **Sector**: Agriculture.
- **Case**: An AI-based crop advisory system in Kenya failed to account for local climatic conditions, leading to crop failures and financial losses for farmers. The system's limitations highlighted the importance of localizing AI systems for diverse environments.

South America

- **Sector**: Banking.
- **Case**: A credit scoring system in Brazil excluded rural populations due to incomplete data, exacerbating financial exclusion and inequality. Advocacy groups called for more inclusive data practices.

Conclusion

The risks and impacts of not utilizing Responsible AI practices are far-reaching, affecting industries and societies globally. From financial penalties to ethical dilemmas, the consequences of neglecting RAI can be devastating. By learning from past failures and adopting proactive measures, organizations can mitigate these risks, build trust, and ensure that AI technologies are leveraged responsibly for the benefit of all. This chapter underscores the necessity of integrating RAI into AI strategies to foster sustainable and ethical innovation.

Chapter 11

Lifecycle of Responsible AI

Introduction

The Responsible AI lifecycle is a structured framework that ensures AI systems are designed, developed, deployed, and maintained ethically and sustainably. Each stage of this lifecycle focuses on embedding principles of fairness, transparency, accountability, privacy, and inclusivity into the AI pipeline. By following this lifecycle, organizations can address potential risks and align their AI systems with societal values, regulatory requirements, and business goals. This chapter explores the stages of the Responsible AI lifecycle, the activities involved at each stage, and the best practices to ensure success.

1. Ideation and Problem Definition

Objective:

Define the purpose of the AI system while considering ethical implications and societal impact.

Key Activities:

- Stakeholder Engagement:
  - Involve diverse stakeholders, such as domain experts, regulators, customers, and communities, to understand the broader implications of the AI system.
  - Collaborate to identify societal, economic, and organizational goals, ensuring the AI system aligns with shared values.
  - Address potential conflicts of interest and include underrepresented groups in the decision-making process.

- Impact Assessment:
  - Conduct a comprehensive ethical impact assessment to evaluate potential benefits, risks, and unintended consequences.
  - Use scenario planning to explore the outcomes for different user groups under various conditions.
  - Prioritize risks based on their likelihood and severity, and document mitigation strategies.
- Define Success Metrics:
  - Develop KPIs that align with Responsible AI principles, such as fairness, transparency, and accountability.
  - Ensure metrics capture technical performance, user satisfaction, and societal impact.
  - Establish measurable goals to track progress and guide system development.

2. Data Collection and Preprocessing

Objective:

Ensure that data used for training AI systems is representative, unbiased, and compliant with privacy regulations.

Key Activities:

- Data Sourcing:
  - Collect data from a diverse set of reliable sources to capture the heterogeneity of the target population.
  - Avoid over-reliance on historical or biased datasets that may perpetuate existing inequities.
  - Engage data custodians to ensure data quality and provenance.
- Bias Detection and Mitigation:
  - Use automated tools and statistical methods to identify patterns of bias in the dataset.
  - Compare distributions of key attributes (e.g., gender, ethnicity) across different groups to ensure balance.
  - Implement strategies such as oversampling underrepresented groups or reweighting data to mitigate biases.
- Privacy Preservation:
  - Employ anonymization techniques, such as removing personally identifiable information (PII), to protect user privacy.

    - Implement differential privacy to add noise to data while preserving aggregate insights.
    - Ensure adherence to data governance frameworks like GDPR, HIPAA, or CCPA.
- Data Quality Assurance:
    - Conduct rigorous validation checks to identify and address anomalies, missing values, or duplicates.
    - Implement tools for real-time data monitoring to ensure continuous quality.
    - Document data preprocessing steps for transparency and reproducibility.

3. Model Development

Objective:

Design and train AI models that adhere to ethical standards and Responsible AI principles.

Key Activities:

- Algorithm Selection:
    - Choose algorithms that are compatible with fairness, interpretability, and efficiency goals.
    - Evaluate the trade-offs between accuracy and ethical considerations, such as privacy and explainability.
    - Test multiple algorithms to identify the best fit for the specific application and context.
- Bias and Fairness Testing:
    - Implement fairness metrics, such as demographic parity or equalized odds, to evaluate model performance across groups.
    - Use tools like IBM AI Fairness 360 to systematically identify and mitigate model biases.
    - Continuously refine training data and model parameters to address disparities.
- Explainability Integration:
    - Incorporate explainability frameworks, such as SHAP and LIME, to clarify model decision-making processes.
    - Design model outputs to include human-readable explanations tailored to diverse stakeholders.

    - Test explainability features with end-users to ensure they meet their needs.
- Adversarial Testing:
    - Simulate adversarial attacks to evaluate model robustness against malicious inputs.
    - Identify vulnerabilities in the model and develop safeguards to mitigate risks.
    - Regularly test models with updated adversarial scenarios to adapt to evolving threats.

4. Deployment and Integration

Objective:

Deploy AI systems in real-world environments while ensuring they operate ethically and effectively.

Key Activities:

- Ethical Review:
    - Conduct a comprehensive review of the AI system to ensure compliance with ethical guidelines and regulatory requirements.
    - Validate that all identified risks during development have been addressed and mitigated.
    - Document the ethical review process for accountability and stakeholder trust.
- Secure Deployment:
    - Implement robust security measures, such as encryption, access controls, and secure authentication, to protect the system and data.
    - Use containerization and virtualization to ensure safe and consistent deployment.
    - Establish protocols for disaster recovery and failover to minimize disruptions.
- User Training and Education:
    - Provide training sessions for end-users, covering the AI system's capabilities, limitations, and ethical implications.
    - Develop user-friendly documentation and FAQs to support system adoption.
    - Offer ongoing support to address user concerns and gather feedback.

- Feedback Mechanisms:
  - Establish structured feedback channels, such as surveys or feedback forms, to collect user experiences and suggestions.
  - Use feedback to identify areas for improvement and prioritize updates.
  - Ensure feedback loops are inclusive, capturing diverse perspectives from all stakeholders.

5. Monitoring and Maintenance

Objective:

Continuously monitor AI systems to ensure ongoing compliance with Responsible AI principles and adapt to changing conditions.

Key Activities:

- Performance Monitoring:
  - Use KPIs, such as accuracy, fairness, and efficiency, to track system performance over time.
  - Set thresholds for acceptable performance and trigger alerts for deviations.
  - Regularly review performance reports to identify trends and anomalies.
- Bias Audits:
  - Conduct periodic audits to detect emerging biases in AI outputs.
  - Use tools like Google's What-If Tool to evaluate fairness across demographic groups.
  - Document audit findings and implement corrective actions promptly.
- Regulatory Compliance Checks:
  - Stay updated on evolving regulations and ensure AI systems adhere to new requirements.
  - Conduct compliance assessments to identify gaps and address them proactively.
  - Maintain detailed records of compliance activities for transparency.
- Incident Management:
  - Develop protocols for identifying, reporting, and resolving ethical or technical incidents.
  - Investigate incidents thoroughly and implement measures to prevent recurrence.
  - Communicate incident outcomes transparently to stakeholders.

- Model Updates and Retraining:
    - Incorporate new data to improve model performance and adapt to changing environments.
    - Validate updates rigorously to ensure they align with ethical and operational standards.
    - Schedule regular retraining cycles to maintain system relevance and accuracy.

6. Decommissioning and Retirement

Objective:

Safely retire AI systems that are no longer effective or aligned with organizational goals and Responsible AI principles.

Key Activities:

- Impact Assessment:
    - Evaluate the impact of decommissioning on stakeholders, including users, employees, and customers.
    - Develop a transition plan to minimize disruptions and ensure continuity of services.
    - Communicate changes clearly to all affected parties.

- Data and Model Disposal:
    - Ensure secure deletion of sensitive data and models to protect user privacy.
    - Document decommissioning activities to maintain accountability and compliance.
    - Archive non-sensitive data for potential future use, following ethical guidelines.
- Knowledge Transfer:
    - Capture lessons learned and best practices from the decommissioned AI system.
    - Share insights with internal teams and external stakeholders to inform future projects.
    - Develop case studies to highlight successes and challenges for organizational learning.

Best Practices for the Responsible AI Lifecycle

1. Stakeholder Involvement:
   - Engage diverse stakeholders, including technical teams, end-users, regulators, and community representatives, at every stage of the lifecycle. This ensures that diverse perspectives are considered, resulting in a more inclusive AI system. By incorporating feedback and concerns from all stakeholders, organizations can preemptively address ethical challenges and build trust.
2. Regular Audits:
   - Conduct audits periodically to identify risks, such as biases or security vulnerabilities, throughout the AI lifecycle. Audits should leverage automated tools and frameworks to ensure thoroughness and consistency. Documentation of audit findings and corrective actions taken enhances transparency and accountability.
3. Documentation and Transparency:
   - Maintain comprehensive documentation of data sources, preprocessing methods, algorithm choices, and system performance metrics. Transparency in these processes builds trust with stakeholders and simplifies compliance with regulations. Detailed records also facilitate troubleshooting and system improvements.
4. Continuous Improvement:
   - Treat the Responsible AI lifecycle as an iterative process, regularly updating models and processes based on new data, technological advancements, and user feedback. Continuous learning ensures that AI systems remain relevant, effective, and aligned with ethical standards in dynamic environments.
5. Ethical Governance:
   - Establish governance structures, such as ethics committees or Responsible AI task forces, to oversee the lifecycle. These bodies should define ethical guidelines, monitor compliance, and make decisions on complex ethical dilemmas. Governance ensures accountability and reinforces an organization's commitment to ethical AI practices.

Process Flow for the Responsible AI Lifecycle

1. Ideation and Problem Definition
2. Data Collection and Preprocessing
3. Model Development

4. Deployment and Integration
5. Monitoring and Maintenance
6. Decommissioning and Retirement

Conclusion

The lifecycle of Responsible AI provides a comprehensive framework for designing, developing, deploying, and maintaining AI systems ethically and sustainably. By embedding Responsible AI principles into each stage, organizations can ensure that their AI systems align with societal values, mitigate risks, and create long-term value for stakeholders. This structured approach empowers organizations to innovate responsibly, fostering trust and accountability in the age of AI.

Chapter 12

Planning for Compliance: Approach, Framework, and Methodology

Introduction

Responsible AI regulations require enterprises to adopt structured approaches to ensure compliance while fostering ethical innovation. Planning for regulatory compliance involves creating a comprehensive framework, defining clear methodologies, and implementing robust processes that align with legal and ethical standards. This chapter explores the approaches, frameworks, and methodologies enterprises use to effectively plan for compliance with Responsible AI regulations.

The Approach to Compliance Planning

1. Understanding Regulatory Landscape
    - Action: Analyze applicable regulations and standards such as GDPR, AIDA, UK Equality Act, or industry-specific guidelines.
        - Conduct in-depth research on regulations relevant to the enterprise's operational domain: Research the specific legal and ethical requirements that apply to your industry, operational geography, and type of AI systems being deployed. This includes understanding data privacy laws, anti-discrimination acts, and sector-specific rules.
        - Collaborate with legal and compliance teams to interpret and contextualize regulatory requirements: Work closely with legal experts to break down complex regulatory language into actionable steps. This ensures that compliance measures are both accurate and aligned with business operations.

  - Stay updated on evolving regulations through regular engagement with regulatory bodies and industry forums: Participate in webinars, forums, and updates from regulatory agencies to keep track of changes or additions to AI-related laws. Proactive engagement prevents non-compliance due to outdated practices.
- Outcome: Develop a clear understanding of legal obligations, ethical considerations, and potential risks associated with non-compliance. This helps organizations create a proactive compliance strategy.

2. Stakeholder Involvement
   - Action: Engage stakeholders including legal teams, data scientists, compliance officers, and external experts.
     - Identify key stakeholders responsible for AI governance, risk management, and implementation: Map the organizational structure to identify who owns compliance-related responsibilities. This includes leaders from legal, IT, data science, and risk departments.
     - Facilitate workshops and discussions to align goals and understand perspectives: Organize collaborative sessions to bring different teams together. This helps in aligning compliance objectives with technical and business strategies while addressing any concerns or gaps.
     - Include external experts for unbiased assessments and additional insights: External consultants can provide fresh perspectives and help validate internal approaches to compliance. Their expertise can also be valuable in understanding emerging global standards.
   - **Outcome:** Ensure comprehensive input, buy-in from all departments, and alignment across organizational units. This fosters a unified compliance strategy that is practical and inclusive of diverse perspectives.
3. Risk Assessment and Prioritization
   - Action: Identify potential risks associated with AI systems, prioritize them based on impact and likelihood.
     - Conduct scenario analysis to understand potential risks: Use simulations to explore "what-if" scenarios where AI systems might fail, behave unexpectedly, or lead to undesirable outcomes. This helps identify vulnerabilities.
     - Use risk quantification frameworks to evaluate and rank risks: Adopt quantitative tools to assign scores to risks based on severity and likelihood. Prioritization ensures that critical risks are addressed first.

    - Document mitigation strategies for each identified risk: Create actionable plans to address each risk, including preventive measures and contingency actions in case risks materialize.
  - Outcome: Develop risk management strategies tailored to organizational needs, ensuring proactive handling of potential challenges. This minimizes disruptions and builds confidence in AI systems.
4. Integration with Business Strategy
  - Action: Align compliance efforts with organizational goals and AI strategies.
    - Define the strategic value of AI systems in achieving business objectives: Clearly articulate how AI systems contribute to organizational goals such as efficiency, innovation, or customer experience. This helps justify compliance investments.
    - Ensure compliance measures complement innovation strategies without stifling creativity: Design compliance frameworks that allow room for experimentation and innovation, balancing regulatory adherence with the need for agile development.
    - Incorporate compliance goals into strategic AI roadmaps and KPIs: Embed compliance milestones into broader AI implementation plans. Use performance indicators to track adherence alongside other business outcomes.
  - **Outcome:** Create synergy between innovation and regulatory adherence, enhancing organizational competitiveness. This ensures that compliance supports business growth rather than hindering it.

The Framework for Responsible AI Compliance

1. Governance Structure
  - Components:
    - Appoint an AI Ethics Officer to oversee compliance and governance: Assign a dedicated individual to spearhead efforts in defining, implementing, and monitoring ethical practices and compliance standards. This role ensures accountability and serves as a central point for decision-making.
    - Establish an AI Governance Committee comprising diverse stakeholders: Form a multidisciplinary team with representatives from legal, technical, operational, and ethical domains to ensure a

holistic approach to governance. This committee helps in balancing technical feasibility with ethical imperatives.

- Define roles and responsibilities for compliance, risk management, and policy enforcement: Clearly articulate duties for all team members involved in AI governance, ensuring that responsibilities for monitoring, reporting, and decision-making are well-distributed and documented.

- Purpose: Ensure accountability and oversight through structured governance mechanisms that enable proactive regulatory adherence. A strong governance framework provides clarity, promotes collaboration, and supports organizational commitment to ethical AI practices.

2. Policy Development
   - Components:
     - Develop an AI ethics policy that defines organizational values and compliance principles: Create a document that outlines the organization's commitment to fairness, transparency, and accountability in AI development and deployment.
     - Create detailed protocols for data usage, model training, validation, and deployment: Establish standardized procedures to guide teams in managing data, developing models, and deploying systems in compliance with ethical and legal requirements.
     - Define escalation paths for reporting non-compliance or ethical concerns: Designate channels and processes for raising issues, ensuring swift and transparent handling of concerns related to AI practices.
   - Purpose: Provide a clear roadmap for ethical AI practices, fostering consistency and accountability across operations. Policies set the foundation for decision-making and align all stakeholders with organizational values and regulatory requirements.
3. Risk Management
   - Components:
     - Conduct AI impact assessments to identify ethical, legal, and operational risks: Evaluate the potential consequences of deploying AI systems, considering factors like bias, unintended outcomes, and data security.
     - Implement mitigation strategies such as algorithmic fairness checks, data anonymization, and bias correction measures: Deploy tools and

techniques to minimize risks, ensuring AI systems operate in a fair and secure manner.

- Establish risk monitoring mechanisms for ongoing evaluation: Use monitoring systems to continuously track and address emerging risks, adapting mitigation strategies as needed.

- Purpose: Minimize risks related to bias, privacy, safety, and reputation, ensuring secure and ethical AI deployment. A proactive risk management approach safeguards the organization and its stakeholders from unforeseen challenges.

4. Transparency Mechanisms
   - Components:
     - Document decision-making processes, including data sources, algorithms used, and outcomes: Maintain thorough records to allow stakeholders to trace decisions back to their origins, ensuring accountability and clarity.
     - Develop explainability tools to provide insights into AI decisions for stakeholders and regulators: Use technologies and frameworks to make AI decisions interpretable, enhancing trust among users and regulators.
     - Publish transparency reports for high-stakes AI systems: Share publicly accessible documents detailing the system's purpose, methods, and results to demonstrate accountability and commitment to ethical practices.
   - Purpose: Foster trust and accountability by making AI operations comprehensible to internal and external audiences. Transparency mechanisms ensure stakeholders understand and trust the systems they interact with.
5. Continuous Monitoring and Auditing
   - Components:
     - Deploy monitoring tools for real-time compliance checks, covering data usage, model behavior, and outcomes: Implement software solutions to track AI systems' operations in real-time, ensuring they adhere to compliance and performance standards.
     - Schedule regular audits to assess adherence to policies and identify improvement areas: Conduct periodic evaluations to measure compliance effectiveness, identify gaps, and recommend corrective actions.

  - Establish feedback loops for incorporating audit findings into compliance frameworks: Use audit results to refine policies, enhance practices, and address weaknesses, creating a cycle of continuous improvement.
- Purpose: Ensure ongoing adherence to regulations, enabling proactive identification and resolution of compliance gaps. Regular monitoring and audits strengthen trust and ensure sustained regulatory compliance.

6. Training and Education
   - Components:
     - Conduct regular training sessions on Responsible AI principles and regulatory requirements for all stakeholders: Organize educational programs to keep employees informed about compliance standards and ethical practices.
     - Develop role-specific training modules for technical teams, legal experts, and decision-makers: Tailor training to the needs of different groups, ensuring relevance and practical application.
     - Create awareness campaigns to emphasize the importance of ethical AI: Use internal communication channels and workshops to foster a culture that prioritizes responsible AI practices.
   - **Purpose:** Equip employees with knowledge and skills to comply with regulations, fostering a culture of responsibility and awareness. Comprehensive education ensures that everyone in the organization contributes to achieving compliance goals.

The Methodology for Compliance Planning

1. Assessment Phase
   - Steps:
     - Identify applicable regulations relevant to the organization's operations and jurisdictions: Map out the regulatory landscape, including regional, national, and industry-specific guidelines that impact the organization's AI systems. Prioritize those with immediate or significant implications.
     - Conduct gap analysis to evaluate current compliance status and identify areas needing improvement: Review existing AI systems, policies, and processes to pinpoint deviations from regulatory requirements. Use findings to identify areas for enhancement or immediate action.

- Tools:
    - Use regulatory checklists to ensure comprehensive coverage: Employ structured checklists to validate alignment with all necessary legal and ethical standards, ensuring no aspect is overlooked.
    - Leverage AI risk assessment frameworks to identify and prioritize potential risks: Apply established frameworks to systematically evaluate vulnerabilities and determine where resources should be allocated to mitigate risks effectively.

2. Design Phase
    - Steps:
        - Develop detailed policies, frameworks, and protocols tailored to organizational and regulatory needs: Craft clear, actionable documents outlining ethical guidelines, data handling practices, and governance models specific to the organization's operations and industry.
        - Create a compliance roadmap with timelines, milestones, and responsibilities: Design a phased plan to implement compliance measures, including short-term actions and long-term objectives. Assign ownership to ensure accountability.
    - Tools:
        - Use policy templates for consistency and efficiency: Start with customizable templates to create comprehensive, uniform policies that meet regulatory and organizational standards.
        - Adopt compliance management platforms for streamlined implementation: Use technology to manage, track, and report on compliance initiatives, ensuring all aspects are executed and monitored effectively.
3. Implementation Phase
    - Steps:
        - Deploy governance structures and processes, ensuring accountability at all levels: Establish and operationalize committees, roles, and workflows to enforce compliance measures and address challenges promptly.
        - Integrate compliance measures into all stages of the AI lifecycle, from data collection to model deployment: Embed regulatory adherence

into every phase, ensuring consistent application from development to operationalization.

- Tools:
    - Utilize workflow automation tools to enhance efficiency: Automate repetitive tasks, such as documentation or reporting, to free resources for strategic compliance activities.
    - Employ explainability and fairness toolkits to support transparent and ethical AI operations: Use tools to validate model outputs, ensuring they align with fairness standards and are interpretable for stakeholders.

4. Monitoring and Improvement Phase
    - Steps:
        - Perform regular audits to assess adherence and effectiveness of compliance measures: Schedule and conduct audits to evaluate system performance, detect non-compliance, and identify areas for improvement.
        - Update policies, frameworks, and protocols based on findings and new regulatory developments: Revise documents and processes in response to audit outcomes and evolving regulatory landscapes to maintain alignment.
    - Tools:
        - Deploy advanced audit tools for detailed evaluations: Leverage sophisticated software to conduct in-depth reviews of data usage, model behavior, and overall compliance.
        - Use AI monitoring dashboards for real-time compliance insights: Implement dashboards that provide continuous visibility into compliance metrics, enabling timely decision-making and corrective action.

Checklist for Responsible AI Compliance Planning

Phase	Key Actions	Tools and Resources
Assessment Phase	Identify applicable regulations	Use regulatory checklists to map laws and standards that apply to the organization's operations and jurisdictions. Engage legal experts for accurate interpretations and actionable insights.

	Conduct gap analysis	Leverage risk assessment frameworks to compare existing processes with regulatory requirements and identify areas for improvement. Document gaps and prioritize them based on risk impact.
Design Phase	Develop policies and frameworks	Use policy templates to establish consistent and comprehensive guidelines for ethical AI practices, addressing data usage, model governance, and accountability. Compliance management platforms ensure policy alignment and dissemination.
	Create a compliance roadmap	Use project management tools to design a phased implementation plan with clear timelines, milestones, and assigned responsibilities. This ensures that compliance initiatives are tracked and executed efficiently.
Implementation Phase	Deploy governance structures and processes	Utilize workflow automation tools to operationalize compliance frameworks. Assign roles and establish accountability at all organizational levels to manage ongoing compliance efforts.
	Integrate compliance measures into AI lifecycle	Employ explainability and fairness toolkits to embed regulatory adherence into every stage of AI development, from data collection to deployment. Validate systems to ensure they meet ethical and legal standards.
Monitoring Phase	Perform regular audits	Use advanced audit tools to systematically evaluate AI systems for adherence to policies, identifying weaknesses and areas for refinement. Regular audits ensure sustained compliance and foster continuous improvement.
	Update policies based on	Continuous learning platforms can help track evolving regulations, allowing

	regulatory changes	organizations to revise policies, frameworks, and practices proactively. This keeps compliance measures up to date and relevant.

Conclusion

Planning for compliance with Responsible AI regulations is a multifaceted process that requires a strategic approach, robust frameworks, and effective methodologies. By adopting the outlined steps and leveraging appropriate tools, enterprises can align their AI initiatives with regulatory requirements, mitigate risks, and build trust among stakeholders. This chapter serves as a guide for organizations to navigate the complexities of Responsible AI compliance and foster ethical innovation.

Chapter 13

Design Principles for a Responsible AI Application

Introduction

Designing Responsible AI applications requires a thoughtful approach that integrates ethical principles, technical excellence, and user-centric considerations. This chapter outlines key design principles for building AI systems that are fair, accountable, transparent, and aligned with societal and organizational values. These principles serve as a foundation for ensuring that AI applications operate responsibly while fostering trust and adoption.

Core Design Principles for Responsible AI Applications

1. Fairness

   Ensuring fairness in AI systems involves minimizing biases and promoting equitable outcomes for all users.

   - Eliminate Bias in Data:
     - Diverse and representative datasets should be used to train AI models, ensuring inclusion of all demographic groups.
     - Conduct regular audits to identify and address imbalances or biases in data collection, preparation, and usage.
     - Implement techniques such as re-sampling, re-weighting, or algorithmic corrections to mitigate data biases effectively.
   - Equitable Algorithm Design:
     - Apply fairness-aware algorithms that are designed to avoid favoring any specific group unfairly.
     - Perform cross-group analysis to validate that model outputs are consistent across various demographics, such as age, gender, and ethnicity.

    - Regularly test and refine models to ensure they do not unintentionally introduce new biases during operation.
- Mitigation Strategies:
    - Use tools like IBM's AI Fairness 360 to assess and correct unfair outcomes in AI systems.
    - Establish fallback mechanisms that allow human oversight and intervention to address identified unfair decisions in real-time.
    - Document bias mitigation steps transparently to facilitate accountability and continual improvement.

2. Transparency

Transparency builds trust by making AI systems' operations understandable to users and stakeholders.

- Explainability:
    - Ensure that AI systems are designed with interpretable models or augmented with explainability tools for black-box models.
    - Provide stakeholders with clear and accessible explanations about how decisions are made, especially in high-stakes scenarios.
    - Use visualization techniques and simplified language to help non-technical users understand the rationale behind AI outputs.
- Documentation:
    - Maintain detailed records of all data sources, preprocessing methods, algorithm choices, and model configurations.
    - Create user-friendly transparency reports for high-impact applications, summarizing the AI's design, purpose, and potential risks.
    - Use a version-controlled repository to track changes and updates made to AI systems over time.
- User Communication:
    - Design intuitive user interfaces that provide clear feedback on how AI decisions are generated and offer explanations where necessary.
    - Implement features that allow users to query and challenge AI decisions, fostering interactive transparency.
    - Proactively disclose AI capabilities and limitations to set realistic user expectations.

3. Accountability

Establishing accountability ensures that organizations and developers are responsible for the outcomes of their AI systems.

- Defined Roles and Responsibilities:
  - Assign accountability to specific roles, such as an AI Ethics Officer, responsible for overseeing the ethical deployment of AI systems.
  - Establish governance structures that distribute responsibilities across legal, technical, and operational teams.
  - Ensure that roles are clearly documented and communicated to all stakeholders involved in the AI lifecycle.
- Error Handling and Reporting:
  - Develop robust mechanisms for identifying and reporting errors or unintended consequences during AI operations.
  - Set up a clear process for rectifying issues, including escalation paths and timelines for resolution.
  - Encourage external reporting channels to allow stakeholders and users to raise concerns about AI performance or behavior.
- Compliance Frameworks:
  - Align AI applications with regional and global regulations, including GDPR, AIDA, or industry-specific guidelines.
  - Regularly audit systems for compliance and ensure adherence to ethical principles such as fairness and transparency.
  - Establish penalties for non-compliance to reinforce organizational accountability.

4. Privacy and Security

   Protecting user data is critical for maintaining trust and complying with data protection regulations.

   - Data Minimization:
     - Collect only the data that is essential for the functionality of the AI application, reducing unnecessary risks.
     - Anonymize sensitive data wherever possible to protect user identities and prevent unauthorized linkages.
     - Periodically review and delete data that is no longer needed for the AI system.
   - Secure Data Storage and Processing:
     - Use advanced encryption methods to protect data during transmission and storage, ensuring confidentiality.
     - Implement secure processing pipelines that prevent unauthorized access or tampering.

    - Regularly conduct penetration testing and vulnerability assessments to identify and fix security gaps.
- User Consent and Control:
    - Provide clear opt-in mechanisms that explain what data is collected and how it will be used.
    - Allow users to access, modify, or delete their data easily through user-friendly interfaces.
    - Notify users promptly in case of data breaches and provide transparent mitigation steps.

5. Inclusivity

   Designing inclusive AI systems ensures that diverse user needs and perspectives are considered.

   - Accessibility:
       - Incorporate features like text-to-speech, adjustable interfaces, and visual aids to make AI applications usable by people with disabilities.
       - Test applications with diverse user groups to identify and address accessibility barriers.
       - Follow recognized accessibility standards, such as WCAG (Web Content Accessibility Guidelines), in AI design.
   - Cultural Sensitivity:
       - Adapt AI systems to respect linguistic, cultural, and societal nuances in different regions.
       - Avoid making assumptions or generalizations that could alienate specific user groups.
       - Engage local stakeholders to provide input during the development process.
   - Stakeholder Engagement:
       - Involve a diverse group of stakeholders, including underrepresented communities, in the design and testing phases.
       - Use feedback loops to continuously incorporate stakeholder input into iterative design improvements.
       - Conduct regular surveys or focus groups to ensure inclusivity goals are being met.

6. Robustness and Reliability

   AI systems must be resilient to adversarial attacks and operate consistently under varying conditions.

- Adversarial Testing:
  - Simulate adversarial scenarios to identify vulnerabilities in the AI system and develop mitigation strategies.
  - Use adversarial training techniques to improve the system's ability to handle malicious inputs.
  - Partner with external security experts to validate the robustness of the AI application.
- Resilience to Uncertainty:
  - Design models to handle incomplete or ambiguous data inputs gracefully without compromising output quality.
  - Implement fail-safe mechanisms that allow systems to revert to predefined defaults in case of uncertainty.
  - Monitor edge cases and refine algorithms to handle unexpected inputs effectively.
- Continuous Monitoring:
  - Deploy real-time monitoring systems to track performance metrics and detect anomalies during operation.
  - Set thresholds and alerts to notify operators of potential issues requiring immediate attention.
  - Use monitoring insights to inform system updates and future improvements.

7. Sustainability

   Minimizing the environmental impact of AI applications contributes to their overall responsibility.

- Energy Efficiency:
  - Optimize AI models to reduce computational resource usage without sacrificing performance.
  - Use energy-efficient hardware and prioritize renewable energy sources for data centers.
  - Adopt model compression techniques to lower energy consumption during training and inference.
- Resource Allocation:
  - Avoid over-provisioning computational resources and balance workloads efficiently.
  - Use lightweight models where possible to reduce the carbon footprint of AI deployments.

  - Encourage sharing of pre-trained models to minimize redundant resource utilization.
- Lifecycle Management:
  - Design hardware and software for long-term usability and ease of maintenance.
  - Develop strategies for recycling and repurposing hardware components at the end of their lifecycle.
  - Regularly update software to ensure compatibility with sustainable technologies.

Implementing Design Principles in Practice

Step 1: Define Ethical Goals and Objectives

- Align the AI application's goals with organizational values and ethical principles. Ensure that the desired outcomes are consistent with the organization's mission, vision, and stakeholder expectations.
- Conduct ethical risk assessments to identify potential challenges and define mitigation strategies. Analyze how the AI system could inadvertently cause harm or fail to align with ethical standards, and develop proactive measures to address these risks.

Step 2: Incorporate Principles into Development Workflows

- Integrate fairness, transparency, and accountability checks at every stage of the development lifecycle. Ensure that these principles are embedded in data collection, model training, evaluation, and deployment processes.
- Use tools like Google's What-If Tool or Microsoft's Responsible AI Dashboard to monitor compliance with Responsible AI principles. Leverage these tools to identify issues, test scenarios, and validate that the system adheres to ethical guidelines.

Step 3: Engage Stakeholders Throughout the Process

- Involve diverse perspectives, including external advisors and user representatives, during design and validation. Include voices from different demographics and expertise areas to ensure the AI system meets varied needs and minimizes biases.
- Conduct user studies to ensure that AI systems meet the needs and expectations of various demographics. Gather feedback on system usability, fairness, and transparency to refine the design and improve user satisfaction.

Step 4: Monitor and Iterate Post-Deployment

- Establish mechanisms for real-time monitoring and feedback to track AI system performance in production. Implement dashboards and alert systems to identify anomalies or deviations from expected behavior.
- Regularly update models to address new challenges, data shifts, and evolving regulatory standards. Continuously improve the AI system based on user feedback, performance metrics, and emerging industry practices.

Real-World Examples of Responsible AI Design

Healthcare: Bias-Free Diagnostic Tools

- A leading AI healthcare company audited its diagnostic algorithms to ensure consistent accuracy across demographic groups, addressing historical biases in healthcare datasets.
- They implemented real-time fairness monitoring to prevent disparities in diagnostic outcomes.

Finance: Transparent Credit Scoring Systems

- A financial institution developed an explainable credit scoring model, enabling users to understand why their applications were approved or denied, fostering trust and fairness.
- The system also allowed users to challenge decisions and provided pathways for dispute resolution.

Retail: Inclusive Personalization Engines

- An e-commerce platform used inclusive design practices to create recommendation systems that cater to diverse user preferences, improving customer satisfaction and equity.
- Regular updates to algorithms ensured continued alignment with evolving user demographics and preferences.

Conclusion

Designing Responsible AI applications requires a commitment to ethical principles, user-centered approaches, and robust technical practices. By adhering to these design principles, organizations can create AI systems that are fair, transparent, accountable, and inclusive. Such systems not only build trust and minimize risks but also set the stage for sustainable and impactful innovation. As the field of AI continues to evolve, integrating these principles into the design process will be critical for ensuring that AI technologies serve humanity responsibly.

Chapter 14

Building Ethical AI Workflows

Building ethical AI workflows involves embedding ethical principles into every stage of the AI lifecycle, from conceptualization to deployment and monitoring. By doing so, organizations can ensure that their AI systems are fair, transparent, and aligned with societal values. This chapter provides a comprehensive guide to integrating ethical considerations at each stage of the AI lifecycle, highlighting best practices, tools, and real-world examples.

1. Defining Ethical Objectives

   Ethical AI begins with clear objectives that align with societal norms, regulatory requirements, and organizational values.

   - Identify Core Principles: Start by defining core ethical principles such as fairness, transparency, accountability, and inclusivity. For instance, an AI system designed for hiring must prioritize fairness to avoid discrimination against specific demographic groups.
   - Engage Stakeholders: Involve diverse stakeholders, including ethicists, domain experts, and affected communities, to identify potential ethical risks and opportunities. This collaborative approach ensures a broader perspective on the implications of AI.
   - Set Measurable Goals: Establish measurable goals for ethical performance, such as reducing biases in model predictions or achieving explainability in decision-making. These goals provide a benchmark for evaluating success.

2. Ethical Data Collection and Management
   Data forms the foundation of AI, making ethical data practices essential for responsible AI development.
   - Ensure Informed Consent: Collect data only with the explicit consent of users, providing clear information about how their data will be used. For instance, in healthcare applications, patients should be fully aware of how their medical data contributes to AI systems.
   - Maintain Data Diversity: Ensure datasets are representative of the target population to minimize bias. For example, a facial recognition system must include diverse images across age, gender, and ethnicity to prevent discriminatory outcomes.
   - Anonymize Sensitive Data: Protect user privacy by anonymizing or encrypting sensitive information. Techniques like differential privacy can help balance data utility and confidentiality.
   - Implement Data Audits: Conduct regular audits to identify and rectify imbalances, inaccuracies, or outdated information in datasets. Tools like Pandas and TensorFlow Data Validation (TFDV) can assist in this process.
3. Fair and Transparent Model Development
   Ethical model development involves designing algorithms that are fair, interpretable, and aligned with ethical objectives.
   - Choose Ethical Architectures: Select model architectures that prioritize explainability and fairness. For example, decision trees are often more interpretable than deep neural networks and may be more suitable for applications requiring transparency.
   - Incorporate Fairness Constraints: Use fairness-aware algorithms that incorporate constraints to ensure equitable outcomes. Tools like Fairlearn and AI Fairness 360 can help implement these constraints.
   - Conduct Bias Testing: Evaluate models for potential biases using metrics like demographic parity or equalized odds. This testing should occur iteratively during development to address biases early.
   - Ensure Interpretability: Leverage explainability tools like SHAP or LIME to make model decisions understandable to non-technical stakeholders. For instance, these tools can help explain why a credit application was approved or denied.
4. Ethical Validation and Testing
   Testing ensures that AI systems meet ethical standards before deployment.

- Stress Test for Bias: Simulate real-world scenarios to evaluate model fairness. For example, test an AI hiring tool by submitting resumes from diverse demographic groups and analyzing outcomes.
- Adversarial Testing: Evaluate the model's robustness against malicious inputs. For instance, test a fraud detection system by introducing synthetic fraudulent transactions.
- Evaluate Explainability: Use interpretability metrics to ensure that model decisions can be effectively communicated to end-users. Tools like Google's What-If Tool can help visualize and interpret model behavior.
- Document Findings: Maintain detailed records of all validation and testing processes, including identified issues and implemented fixes. Transparency in testing builds trust among stakeholders.

5. Responsible Deployment

   Deployment is a critical phase where ethical considerations must translate into actionable safeguards.

   - Implement Safeguards: Deploy monitoring systems to track the performance of AI models in real-time. For instance, anomaly detection tools can flag unexpected model behaviors in a financial trading system.
   - Limit Autonomy in High-Stakes Applications: Ensure human oversight in critical decisions, such as medical diagnoses or legal sentencing. HITL (Human-in-the-Loop) systems provide an additional layer of accountability.
   - Transparent Communication: Clearly communicate the role and limitations of AI to users. For example, a chatbot should explicitly state when a human agent is unavailable to manage expectations.
   - Compliance with Regulations: Adhere to relevant legal frameworks, such as GDPR or HIPAA, to ensure data protection and user rights.

6. Monitoring and Continuous Improvement

   Ethical AI requires ongoing evaluation and adaptation to evolving circumstances.

   - Monitor for Drift: Continuously monitor models for performance degradation or emergence of new biases. Tools like MLflow and Azure Monitor can help track these metrics.
   - Incorporate Feedback: Use feedback from users and stakeholders to refine models and address ethical concerns. For example, user feedback on an AI-powered recommendation system can guide improvements in personalization algorithms.

- Regular Audits: Conduct periodic audits to assess compliance with ethical guidelines. These audits should include an evaluation of data practices, model behavior, and real-world impacts.
- Adapt to Changing Norms: Update ethical frameworks and AI systems to align with evolving societal and regulatory standards. For instance, as privacy laws change, incorporate new requirements into data handling processes.

7. Best Practices for Ethical AI Workflows
    - Foster a Culture of Responsibility: Encourage a culture where all team members prioritize ethical considerations. This includes training programs, workshops, and regular discussions on AI ethics.
    - Use Collaborative Tools: Leverage platforms like Jupyter Notebooks and collaboration suites to ensure transparency and documentation throughout the AI lifecycle.
    - Define Accountability Structures: Clearly outline roles and responsibilities for ethical oversight. For instance, designate an AI ethics officer to oversee compliance and address concerns.
    - Leverage Open-Source Tools: Utilize open-source tools like TensorFlow Privacy or Fairlearn to integrate ethical considerations efficiently. These tools offer pre-built functionalities for fairness, explainability, and privacy.
    - Engage with External Experts: Collaborate with external auditors or ethicists to validate the ethical integrity of AI systems. Independent reviews provide an unbiased perspective and enhance credibility.
8. Real-World Examples of Ethical AI Workflows
    - Healthcare Diagnostics: A hospital uses an AI system to predict patient readmission risks. By incorporating anonymized patient data and ensuring human oversight in final decisions, the workflow minimizes biases and prioritizes patient safety.
    - Credit Scoring: A bank implements a fairness-aware credit scoring system, using diverse datasets and explainability tools to ensure equitable decisions. Regular audits and user feedback guide continuous improvements.
    - Content Moderation: A social media platform deploys a HITL system where AI flags potentially harmful content, and human moderators review ambiguous cases. Training moderators on ethical guidelines enhances decision-making quality.

By integrating ethical considerations throughout the AI lifecycle, organizations can build systems that are not only effective but also responsible and trustworthy. These workflows ensure that AI technologies align with societal values, fostering trust and long-term success.

Chapter 15

Human-in-the-Loop (HITL) Systems

Human-in-the-Loop (HITL) systems combine human judgment and AI capabilities to ensure ethical, accurate, and responsible AI operations. By integrating human oversight, HITL systems can address challenges such as bias, errors, and unintended consequences in AI applications, particularly in high-stakes scenarios.

1. Introduction to Human-in-the-Loop (HITL)
   HITL systems involve human participation at critical points in the AI workflow to ensure that decisions align with ethical and operational standards. Unlike fully autonomous systems, HITL frameworks allow humans to intervene, review, and guide AI outputs. This collaboration ensures a balance between AI efficiency and human intuition, making AI applications safer and more reliable. For example, in scenarios like credit approvals, a human reviewer might override an AI decision if contextual nuances indicate a different outcome.
2. Importance of HITL in Responsible AI
   HITL systems are vital for mitigating risks and enhancing accountability in AI systems. Their importance includes:
   - Ethical Oversight: HITL ensures that AI decisions adhere to ethical principles and societal norms, preventing outcomes that may harm individuals or communities. For instance, in hiring processes, human reviewers can ensure that AI tools are not perpetuating biases.
   - Error Correction: By involving humans, errors that AI might overlook, such as subtle contextual factors, can be identified and rectified. This is crucial in domains like healthcare, where incorrect diagnoses can have serious consequences.

- Bias Mitigation: Humans can scrutinize AI outputs for signs of bias and take corrective action, ensuring fairness. This is particularly important in applications like loan approvals, where biased algorithms could unfairly disadvantage specific groups.
- Accountability: Human involvement ensures that there is a clear chain of responsibility for AI decisions, which is critical in high-stakes scenarios like legal sentencing.
- Trust Building: When users see that humans are part of the decision-making process, their confidence in the AI system's reliability and fairness increases.

3. Key Components of HITL Systems

Human Oversight Mechanisms

- Manual Review: Involves humans assessing and approving AI outputs before final decisions are made. For example, in fraud detection, flagged transactions are reviewed by analysts to confirm their validity.
- Decision Escalation: Complex or uncertain cases are automatically escalated to human operators. For instance, autonomous vehicles may transfer control to a human driver during ambiguous traffic situations.
- Feedback Integration: Human feedback is used to retrain AI models and refine their decision-making processes. For example, moderators' decisions in content review platforms are used to improve the AI's ability to detect harmful content.

Intervention Points

- Pre-Processing: Humans validate data inputs to ensure their quality and relevance. For example, data labeling for training a medical diagnostic model may require human oversight to ensure accuracy.
- Model Outputs: Human operators review AI-generated results, such as loan approval decisions, to ensure they align with policy guidelines and ethical standards.
- Post-Decision: Monitoring outcomes after implementation allows for adjustments to the AI system. For instance, performance reviews of AI-assisted hiring tools can identify unintended biases and guide improvements.

Feedback Loops

HITL systems utilize continuous feedback loops where human decisions inform AI improvements. For example, in an e-commerce recommendation system, customer service representatives' feedback on flagged recommendations can help refine the model.

4. Applications of HITL Systems

Healthcare

- Scenario: AI systems assist in diagnosing diseases based on medical imaging. For example, a model might flag potential signs of cancer in X-rays.
- HITL Role: Radiologists review and validate AI-generated diagnoses, providing context-specific insights that the AI might miss.
- Tools: Platforms like Aidoc and Zebra Medical Vision support HITL workflows, ensuring the accuracy of critical medical decisions.

Finance

- Scenario: Fraud detection systems flag suspicious transactions for review, such as unusually large transfers or transactions from unfamiliar locations.
- HITL Role: Human analysts investigate flagged transactions to confirm or dismiss fraud cases, leveraging their expertise to identify patterns that AI might not detect.
- Tools: Tools like FICO and SAS Fraud Management incorporate human oversight to ensure robust fraud detection.

Content Moderation

- Scenario: Social media platforms use AI to detect harmful or inappropriate content, such as hate speech or explicit imagery.
- HITL Role: Moderators review flagged content to determine its context and appropriateness, ensuring nuanced decisions that AI alone cannot make.
- Tools: Systems like Hive Moderation and AWS Rekognition integrate human moderators to enhance content review processes.

Autonomous Vehicles

- Scenario: AI systems manage navigation and object detection in self-driving cars, such as identifying pedestrians and other vehicles.

- HITL Role: Human operators take over control in ambiguous or high-risk situations, such as adverse weather conditions or complex intersections.
- Tools: Simulation environments like CARLA and platforms like Waymo's development toolkit support HITL testing and deployment.

5. Benefits of HITL Systems
   - Enhanced Accuracy: Human oversight ensures that errors in AI predictions are identified and corrected, leading to more reliable outcomes. For instance, a human reviewer can validate anomalies flagged by AI in financial audits.
   - Ethical Safeguards: HITL provides a safety net to ensure decisions are made in line with societal values. This is particularly crucial in applications like legal sentencing, where fairness is paramount.
   - Improved Fairness: Human intervention allows biases in data or algorithms to be identified and mitigated. For example, recruiters can adjust AI-suggested candidate lists to ensure diversity.
   - Regulatory Compliance: HITL systems help organizations meet legal requirements by providing human oversight in areas like data privacy and decision accountability.
   - Operational Efficiency: By focusing human effort on critical decisions, HITL systems optimize workflows and allow AI to handle routine tasks, enhancing overall efficiency.
6. Challenges in Implementing HITL Systems

Scalability

Managing human oversight for large-scale AI applications can be resource-intensive. For instance, moderating millions of social media posts daily requires significant manpower. Selective sampling and automation can help reduce the workload while maintaining oversight quality.

Human Error

Humans may introduce biases or make mistakes during their reviews. Comprehensive training programs and standardized guidelines can minimize these risks, ensuring consistent and fair decision-making.

Cost

HITL systems can be expensive due to the need for skilled human resources. Organizations can mitigate costs by using HITL selectively for high-impact cases while automating routine decisions.

Latency

Human intervention can slow down real-time AI applications. Efficient escalation mechanisms and prioritization of critical decisions can address this challenge, ensuring timely responses.

7. Best Practices for Implementing HITL Systems
    - Define Clear Roles: Clearly outline the responsibilities of human operators and AI systems. For instance, in healthcare, define whether the AI or the doctor has the final say in diagnosis.
    - Provide Training: Equip human operators with the knowledge to interpret AI outputs accurately. Regular workshops and scenario-based training sessions ensure consistent decision-making.
    - Use Selective Sampling: Focus human reviews on high-risk or ambiguous cases. For example, in fraud detection, prioritize transactions flagged as high-risk for human review.
    - Integrate Feedback Loops: Ensure that human feedback is continuously used to improve AI models. In content moderation, decisions made by moderators should be used to refine the AI's ability to flag inappropriate content.
    - Monitor Performance: Regularly evaluate both AI and human components of the HITL system. Use metrics like accuracy, fairness, and user satisfaction to identify areas for improvement.
    - Leverage Technology: Use platforms like AWS SageMaker Ground Truth or Labelbox to streamline data labeling and HITL workflows, enhancing collaboration between humans and AI.
    - Ensure Transparency: Clearly communicate the role of human oversight in AI decision-making to stakeholders. Transparency fosters trust and demonstrates a commitment to ethical practices.
8. Future of HITL Systems

    As AI continues to evolve, HITL systems will remain critical for balancing automation with human judgment. Advances in explainable AI (XAI) and interactive interfaces will enhance the effectiveness of HITL frameworks. Organizations must continue to prioritize human oversight, especially in critical and high-stakes applications, to ensure ethical and responsible AI deployment.

Conclusion

By integrating HITL systems into AI workflows, organizations can enhance accountability, accuracy, and trust in their AI applications. This collaborative approach ensures that AI technologies are aligned with human values and societal expectations.

Chapter 16

Quality Engineering for Responsible AI

Effective testing and validation are critical components of Responsible AI. They ensure that AI systems operate as intended, meet ethical standards, and remain resilient to potential adversarial threats. This chapter explores strategies for bias testing, fairness testing, and adversarial testing, providing actionable insights, tools, scenarios, and best practices for organizations.

1. Importance of Testing and Validation in AI

   Testing and validation serve as the foundation for trust and reliability in AI systems. They are essential for identifying issues such as bias, unfairness, or vulnerabilities before deployment. By thoroughly testing and validating AI systems, organizations can ensure compliance with ethical and regulatory standards, reduce risks associated with errors or unintended consequences, and enhance user confidence in AI applications. Furthermore, robust testing helps organizations proactively address potential adversarial threats, ensuring the system remains secure and resilient in dynamic environments.

2. Bias Testing

   Bias in AI systems can result in unfair outcomes and harm marginalized groups, often stemming from imbalanced datasets or flawed algorithms. Bias testing identifies these issues and ensures that AI systems produce equitable results.

Strategies for Bias Testing:

- Dataset Audits
  Conduct comprehensive audits of training data to identify imbalances or gaps that may lead to biased outcomes. Statistical tools and visualization libraries like Pandas, Matplotlib, and Power BI can help assess demographic representation. For example, in a loan approval dataset, check if specific groups are underrepresented in approved loans.
- Model Outcome Analysis
  Evaluate the outputs of AI models to detect disparities in performance across different demographic groups. Tools like AI Fairness 360 and SHAP can be used to analyze discrepancies. For instance, test if an AI system predicts higher loan rejection rates for specific age groups or genders.
- Bias Detection Tools
  Use platforms like Fairlearn or What-If Tool to automate bias detection. These tools generate metrics like disparate impact and visualize potential sources of bias. A retail recommendation engine could be tested for biased suggestions by analyzing product visibility across different customer segments.
- Iterative Testing

Conduct bias testing throughout the AI lifecycle. For example, test sentiment analysis models after updates to ensure they don't favor specific language structures. Tools like TensorFlow Data Validation (TFDV) can assist in monitoring data changes over time.

Test Data for Bias Testing:

- Use synthetic datasets that mimic real-world scenarios, such as a hiring dataset with equal representation of genders and ethnicities.
- Generate edge cases, such as scenarios where input variables disproportionately affect a specific group.

Best Practices for Bias Testing:

- Collaborate with domain experts who understand the societal implications of biases in the system's application area. For example, a healthcare AI model could involve medical professionals to identify potential patient disparities.
- Involve diverse stakeholders, including data scientists, legal experts, and community representatives, to provide comprehensive oversight.
- Maintain thorough documentation of testing methodologies, findings, and corrective actions to ensure transparency and accountability.

3. Fairness Testing

   Fairness testing ensures that AI systems deliver equitable outcomes for all users, regardless of their demographic or contextual attributes. It is a crucial step to prevent AI from reinforcing societal inequities.

Strategies for Fairness Testing:

- Define Fairness Metrics

  Select appropriate fairness metrics such as demographic parity (equal outcomes across groups) or equalized odds (similar error rates for all groups). Tools like AI Fairness 360 and Fairlearn provide pre-built metrics and visualizations.
- Simulated Scenarios

  Use synthetic data to simulate real-world scenarios, such as customer behavior patterns in different regions. Test recommendation engines by varying input attributes (e.g., income levels) to evaluate their influence on outcomes.
- Stress Testing

  Introduce extreme cases, such as rare combinations of inputs, to evaluate the model's robustness in producing fair results. For instance, test a pricing model's fairness when applied to customers with minimal credit history.
- Cross-Validation

  Split datasets into multiple subsets and validate model fairness across all segments. This ensures that the model's fairness is not limited to specific data portions. For example, test an e-commerce recommendation model's performance on datasets segregated by region or age group.

Test Data for Fairness Testing:

- Generate balanced synthetic datasets representing all demographic attributes equally.
- Use real-world data with annotations for demographic labels to measure outcome disparities.

Best Practices for Fairness Testing:

- Work with interdisciplinary teams, including ethicists and domain experts, to align fairness goals with societal expectations. For example, involve educators when testing AI for student assessments.

- Regularly update fairness benchmarks and metrics to align with evolving legal and societal standards.
- Share fairness testing outcomes with internal and external stakeholders to foster trust and transparency.

4. Adversarial Testing
   Adversarial testing evaluates an AI system's robustness against malicious attacks or inputs designed to exploit its vulnerabilities. This testing is vital for ensuring security and reliability.

Strategies for Adversarial Testing:

- Evasion Testing
  Assess the model's resilience to adversarial inputs, such as subtly altered data designed to cause misclassification. Tools like CleverHans can generate adversarial examples to test robustness. For example, modify an image by adding noise to trick an AI system into mislabeling it.
- Poisoning Attacks
  Simulate attacks where malicious data is introduced during training. Test if the model can identify and resist such data. For example, use altered credit card transaction data to evaluate a fraud detection system.
- Model Inversion
  Evaluate whether sensitive information can be extracted from an AI model. Tools like TensorFlow Privacy can test model defenses against inversion attacks. For example, assess if a facial recognition system can leak identifiable attributes.
- Use Adversarial Tools
  Utilize platforms like Adversarial Robustness Toolbox to create and evaluate attack scenarios. These tools provide frameworks for identifying vulnerabilities and testing defensive strategies.

Test Data for Adversarial Testing:

- Create adversarial examples such as images with pixel-level noise or text with subtle character changes.
- Introduce poisoned datasets with mislabeled or malicious entries to evaluate model defenses.

Best Practices for Adversarial Testing:

- Regularly update testing strategies to counter evolving adversarial techniques and threats.
- Involve cybersecurity experts to create realistic attack scenarios and evaluate model defenses.
- Develop comprehensive incident response plans to mitigate potential impacts of adversarial attacks.

5. Integrating Testing into the AI Lifecycle

Integrating testing and validation into every phase of the AI lifecycle maximizes their effectiveness. During the pre-development phase, organizations should identify potential risks, define testing goals, and select appropriate tools. In the development phase, iterative testing should accompany data preprocessing, feature engineering, and model training. After deployment, real-time monitoring systems should track performance, fairness, and security metrics. Post-deployment evaluations ensure that systems adapt to changing environments and continue meeting ethical standards.

6. Reporting and Transparency

Clear documentation and reporting are essential for ensuring accountability. Organizations should maintain detailed records of testing methodologies, metrics, and findings. Share these outcomes with stakeholders, including internal teams, customers, and regulators, to demonstrate commitment to Responsible AI. Publishing transparency reports can further reinforce the organization's dedication to ethical practices, fostering trust and confidence among users.

7. Challenges and Solutions

Challenges:

- Limited availability of diverse datasets, making bias and fairness testing more difficult.
- Complexity in selecting and interpreting fairness metrics for specific use cases.
- Rapidly evolving techniques for adversarial attacks that can outpace defenses.

Solutions:

- Partner with external organizations to access diverse datasets and expertise.
- Use explainable AI tools to help interpret fairness metrics and provide actionable insights.
- Continuously update testing frameworks and methodologies to address new threats and challenges.

By adopting these comprehensive strategies for bias testing, fairness testing, and adversarial testing, and leveraging appropriate tools and real-world scenarios, organizations can ensure that their AI systems are ethical, secure, and resilient. Rigorous testing and validation not only mitigate risks but also enhance trust and reliability, paving the way for successful and Responsible AI deployments.

Chapter 17

Measuring the Impact of Responsible AI

Introduction

Responsible AI (RAI) is not just a theoretical concept; it represents a commitment to ensuring that Artificial Intelligence (AI) technologies are ethical, transparent, and accountable. However, the adoption of RAI principles requires measurable outcomes to assess its effectiveness and refine implementation strategies. This chapter explores the methodologies, tools, and best practices for measuring the impact of RAI, providing organizations with a framework to evaluate their AI systems comprehensively. By focusing on tangible metrics, organizations can ensure alignment with societal values, regulatory requirements, and business goals.

Key Performance Indicators (KPIs) for Responsible AI (RAI) are essential for quantifying the ethical, operational, and societal impact of AI systems. They provide organizations with measurable benchmarks to track progress, ensure compliance, and foster trust. This chapter outlines essential KPIs for Responsible AI, along with examples and calculation methods, enabling organizations to align their AI systems with ethical standards and business goals effectively.

Essential KPIs for Responsible AI

1. Fairness KPIs
   Definition:
   Fairness KPIs ensure that AI systems produce outcomes that are equitable across diverse user groups. These KPIs help prevent bias and discriminatory practices, ensuring that no demographic is unfairly advantaged or disadvantaged.

Metrics Examples:

- Demographic Parity:
    - **Explanation**: This metric evaluates whether outcomes are equally distributed across different demographic groups.
    - **How to Calculate**: Divide the positive outcome rate of one group by the positive outcome rate of another group.
    - **Example**: If 70% of Group A and 68% of Group B receive approvals, the ratio is 0.97.
    - **Desired Value**: Close to 1 indicates fairness.

Equalized Odds:

- **Explanation**: Measures whether error rates (false positives and false negatives) are consistent across demographic groups.
- **How to Calculate**: Compare false positive and false negative rates for different groups.
- **Example**: If Group A has an FPR of 10% and Group B has 15%, mitigation strategies are required.
- **Desired Value**: Minimal differences in error rates.

Implementation Steps:

1. Conduct demographic analyses on outcomes:
    - Analyze AI outputs for all demographic groups to identify any disparities in outcomes.
    - Use statistical methods to quantify differences in error rates and decision patterns.
    - Regularly update the demographic data to reflect current population distributions.
2. Use fairness auditing tools like IBM AI Fairness 360:
    - Leverage automated tools to assess biases in AI models and datasets.
    - Generate detailed reports highlighting fairness metrics and areas for improvement.
    - Incorporate these insights into the model training and validation processes.
3. Monitor and recalibrate models as needed:
    - Establish periodic checkpoints to evaluate the fairness of AI systems in production.
    - Retrain models with updated or augmented datasets to address identified biases.

- Maintain a log of recalibration efforts for transparency and accountability.

1. Transparency and Explainability KPIs
   Definition:
   Transparency and explainability KPIs measure how well stakeholders can understand and trust AI systems. These metrics are essential for building confidence in the system's decision-making processes.

Metrics Examples:

- Explainability Score:
  - **Explanation**: Reflects the proportion of decisions that are clearly explained to users.
  - **How to Calculate**: Divide the number of explained decisions by the total number of decisions.
  - **Example**: 950 out of 1,000 decisions explained clearly yields a score of 95%.
  - **Desired Value**: Above 90% for high-stakes applications.
- Model Interpretability Index:
  - **Explanation**: Measures the percentage of interpretable models in production.
  - **How to Calculate**: Divide the number of interpretable models by the total number of models.
  - **Example**: 8 interpretable models out of 10 yields an index of 80%.
  - **Desired Value**: Above 80%.

Implementation Steps:

1. Integrate tools like SHAP or LIME:
   - Deploy interpretability tools during the development and testing phases.
   - Use these tools to highlight feature contributions and decision pathways in AI models.
   - Train technical teams to utilize these tools effectively for transparency.
2. Test explanations with end-users to assess clarity:
   - Conduct user surveys or focus groups to evaluate the comprehensibility of explanations.
   - Gather feedback from diverse user groups to ensure inclusivity in understanding.
   - Iterate on explanation methods based on user input to enhance effectiveness.

3. Report interpretability efforts transparently:
    - Publish regular reports detailing how interpretability is measured and improved.
    - Share success stories and challenges with stakeholders to build trust.
    - Include interpretability metrics in internal and external audits.
3. Privacy and Security KPIs
   Definition:
   Privacy and security KPIs ensure that AI systems protect user data and adhere to regulatory standards. These metrics focus on minimizing data breaches and safeguarding sensitive information.

Metrics Examples:

- Data Minimization Rate:
    - **Explanation**: Tracks adherence to the principle of collecting only necessary data.
    - **How to Calculate**: Divide the relevant data collected by the total data collected.
    - **Example**: If 85% of collected data is relevant, the rate is 85%.
    - **Desired Value**: Above 90%.
- Breach Incidence Rate:
    - **Explanation**: Monitors the frequency of security breaches in the system.
    - **How to Calculate**: Divide the number of breaches by total data access events and multiply by 100.
    - **Example**: 2 breaches out of 10,000 data access events result in a rate of 0.02%.
    - **Desired Value**: Below 1%.

Implementation Steps:

1. Conduct regular audits:
    - Perform routine assessments of data handling processes to identify vulnerabilities.
    - Verify adherence to privacy laws such as GDPR and CCPA during audits.
    - Share audit results with stakeholders to promote accountability.

2. Use monitoring tools like AWS CloudTrail:

- Deploy monitoring solutions to track data access and usage patterns in real-time.
- Configure alerts for unauthorized access attempts or unusual activities.
- Maintain logs for investigation and compliance purposes.

3. Implement privacy-preserving techniques:
   - Incorporate techniques like data anonymization and differential privacy.
   - Ensure that these methods are applied consistently across datasets.
   - Test the effectiveness of privacy measures through simulated attack scenarios.
4. Inclusivity and Accessibility KPIs

   Definition:

   Inclusivity and accessibility KPIs measure how well AI systems serve diverse populations, including marginalized and underrepresented groups. These metrics ensure systems are usable and beneficial to all.

Metrics Examples:

Accessibility Compliance Rate:

- **Explanation**: Tracks adherence to accessibility standards like WCAG.
- **How to Calculate**: Divide compliant features by the total features tested, then multiply by 100.
- **Example**: 45 out of 50 features meeting standards yields 90% compliance.
- **Desired Value**: 100%.

Representation Index:

- **Explanation**: Measures the diversity of participants in AI development and testing.
- **How to Calculate**: Divide participants from underrepresented groups by total participants.
- **Example**: 30 out of 100 participants yield an index of 30%.
- **Desired Value**: Above 50%.

Implementation Steps:

1. Engage diverse groups in testing phases:
   - Invite participants from various demographics to test AI systems.
   - Ensure that the testing process captures feedback from underrepresented communities.

   - Use feedback to enhance system inclusivity.
2. Incorporate accessibility features like text-to-speech:
   - Develop features tailored to users with disabilities.
   - Test these features rigorously to ensure functionality and usability.
   - Regularly update accessibility features to align with evolving standards.
3. Review metrics regularly:
   - Establish a schedule for reviewing inclusivity and accessibility metrics.
   - Use reviews to identify gaps and develop targeted improvement plans.
   - Share progress with stakeholders to demonstrate commitment to inclusivity.
5. Environmental Sustainability KPIs

   Definition:

   Environmental sustainability KPIs focus on reducing the ecological footprint of AI systems. These metrics ensure systems are energy-efficient and aligned with sustainability goals.

Metrics Examples:

- Carbon Emissions Rate:
   - **Explanation**: Tracks emissions generated by AI systems during their lifecycle.
   - **How to Calculate**: Divide carbon emissions (kg) by model training hours.
   - **Example**: 500 kg of CO_2 over 100 hours yields 5 kg/hour.
   - **Desired Value**: Lower rates over time.
- Energy Efficiency Score:
   - **Explanation**: Measures computational efficiency in AI systems.
   - **How to Calculate**: Divide useful computations by total energy consumed.
   - **Example**: A score of 85% indicates high efficiency.
   - **Desired Value**: Above 80%.

Implementation Steps:

1. Optimize models for efficiency:
   - Use model compression techniques such as pruning and quantization.
   - Select algorithms that balance accuracy and computational demands.
   - Monitor energy usage during the optimization process.
2. Monitor energy consumption using tools like CodeCarbon:
   - Install tools to track energy usage and carbon emissions in real-time.

    - Use reports to identify energy-intensive processes and reduce their impact.
    - Share results in sustainability reports to align with ESG goals.
3. Transition to renewable energy sources:
    - Partner with green data centers powered by renewable energy.
    - Implement policies to prioritize renewable energy procurement.
    - Measure the impact of renewable energy adoption on overall carbon emissions.

KPI Summary Table

Metric Name	Explanation	How to Calculate	Desired Value
Demographic Parity	Ensures equal outcomes across demographic groups.	Positive Outcome Rate Group A / Group B	Close to 1
Equalized Odds	Measures consistency in error rates across groups.	Compare FPR and FNR for groups	Minimal difference
Explainability Score	Assesses clarity of AI decisions.	Decisions Explained Clearly / Total Decisions	Above 90%
Model Interpretability	Tracks interpretable models in production.	Interpretable Models / Total Models	Above 80%
Data Minimization Rate	Tracks adherence to data minimization.	Relevant Data Collected / Total Data Collected	Above 90%
Breach Incidence Rate	Monitors frequency of security breaches.	(Number of Breaches / Data Access Events) x 100	Below 1%
Accessibility Compliance	Measures adherence to accessibility standards.	(Compliant Features / Total Features Tested) x 100	100%

Representation Index	Tracks diversity in development/testing.	Diverse Participants / Total Participants	Above 50%
Carbon Emissions Rate	Tracks emissions during AI lifecycle.	Carbon Emissions (kg) / Model Training Hours	Lower over time
Energy Efficiency Score	Measures computational efficiency of AI systems.	Useful Computations / Total Energy Consumed	Above 80%

Case Study: Fraud Management Application

To illustrate how KPIs can be applied, consider an AI-powered Fraud Management System used in financial services.

1. Fairness:
   - Ensure that fraud detection rates are consistent across customer demographics, such as age, gender, and geography.
   - Regularly audit error rates (false positives and negatives) to identify and address potential biases.
2. Transparency and Explainability:
   - Use explainability tools to provide clear reasoning for flagged transactions, enabling customers and regulators to understand the decisions.
   - Monitor and report the percentage of decisions explained to stakeholders.
3. Privacy and Security:
   - Implement strict data minimization practices, collecting only necessary transaction details.
   - Use privacy-preserving techniques like encryption to secure sensitive customer data.
4. Inclusivity and Accessibility:
   - Ensure the system interface is accessible to users with disabilities, following WCAG standards.
   - Engage diverse testers to validate system usability across various demographics.
5. Environmental Sustainability:

- Optimize model training to reduce energy consumption, using cloud services powered by renewable energy.
- Monitor the system's carbon footprint and implement strategies to minimize emissions.

By measuring these KPIs, organizations can ensure their Fraud Management System is ethical, effective, and aligned with Responsible AI principles.

Conclusion

Defining and tracking KPIs for Responsible AI ensures that organizations can measure their ethical and operational performance effectively. By focusing on metrics like fairness, transparency, privacy, inclusivity, and sustainability, organizations can build AI systems that are not only innovative but also ethical and trustworthy. These KPIs, supported by advanced tools and clear methodologies, provide a roadmap for responsible AI implementation and continuous improvement.

Chapter 18

Standard Benchmarks, Challenges & Recommendations

The rapid development of Generative AI models, including large language models (LLMs), has intensified the focus on ensuring responsible AI practices. To evaluate these models' performance in alignment with ethical principles and societal norms, standard benchmarks have been established. These benchmarks enable LLM providers to measure and compare their models' adherence to criteria such as fairness, accountability, transparency, privacy, and robustness. This chapter explores the widely recognized benchmarks for responsible AI and how they guide LLM providers in evaluating their models.

1. Key Principles of Responsible AI Benchmarks

Responsible AI benchmarks are grounded in foundational principles that ensure AI technologies align with ethical and societal expectations. These principles include:

- Fairness: AI models should treat all users equitably, avoiding discrimination or bias against any demographic group. This ensures inclusivity and fosters trust in the technology.
- Transparency: Making AI systems understandable to users and stakeholders is essential for accountability. Transparency ensures that decision-making processes can be audited and justified.
- Accountability: Clear delineation of responsibility ensures that any issues arising from AI systems can be addressed effectively, fostering trust among stakeholders.

- Robustness: AI systems must be resilient to errors, adversarial attacks, and unintended behaviors, ensuring reliability in diverse real-world scenarios.
- Privacy: Protecting users' data is critical to maintaining trust and complying with regulations like GDPR and HIPAA.

2. Notable Benchmarks for Responsible AI

a. Ethical and Social Impact Benchmarks

These benchmarks focus on ensuring that AI systems operate ethically and align with societal values:

1. EQUALS Global AI Toolkit:
    - What It Measures: This benchmark assesses how AI technologies promote inclusivity, particularly in terms of gender equality and accessibility. It evaluates the presence of balanced representation in datasets and fairness in outcomes.
    - Desired Value: A fairness score of 0.95 or higher.
    - Why LLMs Fall Short: Many training datasets underrepresent marginalized groups or carry historical biases that perpetuate inequality in outputs.
    - Recommendations:
        - Dataset Curation: Curate datasets that include diverse voices and ensure balanced representation of demographic groups.
        - Bias Mitigation Techniques: Use fairness-aware learning algorithms to identify and reduce biases during training.
        - Continuous Auditing: Regularly audit model outputs for fairness across various demographic and cultural contexts.
2. AI Fairness 360 (IBM):
    - What It Measures: AI Fairness 360 is a suite of metrics and tools that detect and address biases in AI systems. It assesses disparate impact, statistical parity, and other fairness metrics.
    - Desired Value: A fairness score above 0.98.
    - Why LLMs Fall Short: LLMs struggle to maintain fairness due to a lack of demographic-specific context in training data and difficulties in understanding nuanced cultural differences.
    - Recommendations:
        - Fairness Constraints: Incorporate fairness constraints during training to penalize biased outcomes.

- Bias Testing Pipelines: Implement pre-deployment pipelines to test for disparate impacts on sensitive attributes such as race, gender, and age.
- Feedback Loops: Collect feedback from diverse user groups to improve fairness in real-world scenarios.

3. HATECHECK:
   - What It Measures: HATECHECK evaluates models' ability to detect and moderate hate speech. It includes 29 test cases that represent common and subtle forms of hate speech.
   - Desired Value: Precision and recall scores of at least 0.90.
   - Why LLMs Fall Short: LLMs often misclassify hate speech due to insufficient understanding of context or deliberate adversarial inputs.
   - Recommendations:
     - Targeted Training: Train models on curated datasets specifically designed for hate speech contexts, including multilingual examples.
     - Human-In-The-Loop (HITL): Use human moderation to fine-tune models for nuanced scenarios.
     - Adversarial Defense: Develop techniques to recognize adversarial prompts and prevent harmful content generation.
4. GENDER-GAP Dataset:
   - What It Measures: This benchmark evaluates gender bias in LLM-generated text, focusing on stereotypical associations and overgeneralizations.
   - Desired Value: A bias mitigation score above 0.96.
   - Why LLMs Fall Short: Gender biases often stem from historical and imbalanced data used during training, leading to skewed outputs.
   - Recommendations:
     - Counterfactual Data Augmentation: Include counterexamples in training data to offset biased patterns.
     - Fairness-Aware Loss Functions: Incorporate loss functions that explicitly penalize biased outputs.
     - Evaluation Frameworks: Use fairness evaluation frameworks to detect and address stereotyping tendencies.

b. Technical Robustness and Security Benchmarks

These benchmarks ensure models are reliable and resistant to attacks or misuse:

1. Adversarial Robustness Benchmark:
   - What It Measures: This benchmark tests a model's ability to handle adversarial inputs designed to confuse or manipulate its outputs. High robustness scores indicate better resistance.
   - Desired Value: Robustness scores above 0.85.
   - Why LLMs Fall Short: LLMs are vulnerable to adversarial prompts, especially those exploiting weaknesses in language understanding.
   - Recommendations:
     - Adversarial Training: Regularly train models on adversarial inputs to increase resilience.
     - Diverse Testing Scenarios: Include varied adversarial examples to prepare for real-world attacks.
     - Monitoring Tools: Deploy real-time monitoring tools to detect and mitigate adversarial behaviors.
2. SecML:
   - What It Measures: SecML evaluates machine learning models' robustness against adversarial attacks using various testing methods.
   - Desired Value: A robustness score above 0.90.
   - Why LLMs Fall Short: Weak input validation and over-reliance on assumptions make models susceptible to security breaches.
   - Recommendations:
     - Input Filtering: Implement strict input validation mechanisms to block malicious inputs.
     - Ensemble Models: Use ensemble methods to enhance robustness by combining multiple predictions.
3. LAMBADA Benchmark:
   - What It Measures: LAMBADA tests linguistic understanding by requiring models to predict the last word of a sentence in a coherent manner.
   - Desired Value: Accuracy above 0.95.
   - Why LLMs Fall Short: Context retention issues and inadequate fine-tuning often lead to failures in completing sentences accurately.
   - Recommendations:
     - Enhanced Context Mechanisms: Refine model architectures to better retain and utilize context.

    - Targeted Fine-Tuning: Focus on datasets emphasizing linguistic coherence and logical sentence completion.

4. OpenAI Red Teaming:
    - What It Measures: OpenAI's internal benchmarks test models under diverse scenarios, identifying vulnerabilities and unintended behaviors.
    - Desired Value: A performance score above 0.92.
    - Why LLMs Fall Short: Edge cases and ambiguous inputs expose gaps in generalization capabilities.
    - Recommendations:
        - Scenario Expansion: Regularly expand stress-testing scenarios to include diverse and emerging edge cases.
        - External Collaboration: Collaborate with external red-teaming experts to uncover hidden vulnerabilities.

Conclusion

Standard benchmarks for responsible AI play a pivotal role in guiding LLM providers toward ethical, secure, and high-performing AI systems. By adhering to these benchmarks, providers can demonstrate their commitment to societal values, drive innovation, and build trust in AI technologies. As benchmarks continue to evolve, they will remain essential tools for ensuring AI's positive impact on society.

Chapter 19

Responsible AI Approach: Gen AI & Agentic AI

Artificial Intelligence (AI) is transforming industries and societies, yet its rapid development has raised significant ethical, regulatory, and operational challenges. To address these challenges, Responsible AI frameworks ensure AI systems are designed, deployed, and used ethically, transparently, and safely. However, the nature of Responsible AI approaches differs significantly between Generative AI (Gen AI) and Agentic AI due to their distinct functionalities and impacts. This chapter explores these differences in depth.

1. Understanding Generative AI and Agentic AI

Generative AI (Gen AI)

Generative AI models are designed to produce new content, such as text, images, audio, or code, based on patterns learned from training data. Examples include ChatGPT for conversational AI, DALL-E for image generation, and Copilot for code completion.

Key Characteristics of Gen AI:

- Produces outputs based on prompts: Gen AI is reactive, creating responses or outputs based on user-provided inputs, making it highly dependent on the quality of these prompts.
- Outputs can include creative content, recommendations, or summaries: The applications of Gen AI span diverse industries, from marketing to education, emphasizing its flexibility.
- Heavily reliant on the quality and diversity of training data: Biases in training data can directly influence the output, potentially leading to ethical issues.
- Often lacks real-world context or autonomous decision-making capabilities: Gen AI systems do not independently interpret or act on real-world scenarios.

Agentic AI

Agentic AI systems operate autonomously, making decisions and executing actions on behalf of users. These systems are deployed in high-stakes environments, such as autonomous vehicles, robotics, or financial trading platforms.

Key Characteristics of Agentic AI:

- Focused on decision-making and task execution: Agentic AI is goal-driven, designed to perform complex tasks without constant human intervention.
- Operates in real-time environments with dynamic inputs: These systems adapt to changing conditions, requiring advanced decision-making capabilities.
- Carries out actions with minimal human intervention: Once deployed, Agentic AI systems act autonomously within predefined boundaries.
- Includes a higher degree of autonomy and accountability: Agentic AI systems can have significant consequences, necessitating robust safety measures.

2. Key Considerations in Responsible AI

The differences between Gen AI and Agentic AI necessitate tailored Responsible AI approaches. Below, we explore these considerations in detail:

a. Transparency

Generative AI:

- Challenge: Explaining how specific outputs are generated (e.g., hallucinations or unintended biases in responses). Gen AI models often function as black boxes, making it difficult to pinpoint the reasoning behind certain outputs.
- Approach:
  - Maintain transparency about training data sources and limitations: Disclose information about the datasets used to train the model, ensuring users understand the system's context and limitations.
  - Provide detailed documentation on model architecture and tuning methods: Offer clear explanations of how the model functions and the adjustments made during fine-tuning.
  - Implement user-facing disclaimers on the reliability and scope of generated outputs: Warn users about potential inaccuracies or inappropriate outputs.
- Use Case: In education, a Gen AI-powered essay assistant like ChatGPT can help students generate ideas for academic papers. Transparency is essential

here—students should know the AI is trained on publicly available datasets, which might exclude certain academic sources. By explicitly stating its limitations, such as an inability to verify facts or provide citations, students can use the tool responsibly for brainstorming rather than relying on it as a definitive source of truth. This ensures academic integrity while leveraging AI's creative capabilities.

Agentic AI:

- Challenge: Ensuring that autonomous decision-making processes are interpretable, especially in safety-critical environments. Unlike Gen AI, the stakes for explainability are higher in Agentic AI due to its autonomous nature.
- Approach:
  - Incorporate explainability mechanisms (e.g., decision trees or heatmaps) for decisions made by the AI: Use tools and techniques to visualize and communicate the decision pathways.
  - Use interpretable models or supplementary tools to justify actions, particularly in regulated industries such as healthcare or finance: Ensure that regulators and stakeholders can understand and trust the system's decisions.
- Use Case: In healthcare, autonomous diagnostic systems assist doctors by analyzing medical images for early signs of diseases like cancer. For instance, an AI detecting tumors in radiology scans must highlight the specific regions of concern and provide a confidence score for its diagnosis. This ensures doctors can cross-check and validate the AI's recommendation before proceeding with treatment plans, combining human expertise with machine precision to improve patient outcomes.

b. Bias Mitigation

Generative AI:

- Challenge: Outputs may reflect biases present in the training data, perpetuating stereotypes or misinformation. The vast datasets used in training often contain inherent biases that can influence the model's outputs.
- Approach:
  - Use diverse, representative datasets during training: Ensure training data encompasses a wide range of perspectives and reduces the likelihood of reinforcing biases.

  - Continuously audit outputs for bias and harmful content: Implement processes to regularly review and address issues in the model's performance.
  - Implement post-processing filters to detect and correct biased responses: Develop tools to automatically identify and rectify problematic outputs.
- Use Case: In recruitment, a Gen AI tool used to screen resumes can inadvertently favor certain demographics if its training data is biased. For example, if historical hiring data disproportionately favors male candidates, the AI might rank female candidates lower. By incorporating diverse datasets and implementing fairness metrics, the tool can be audited to ensure it evaluates resumes based on skills and qualifications rather than demographic factors, fostering inclusivity in hiring processes.

Agentic AI:

- Challenge: Decisions made autonomously can be disproportionately affected by biased algorithms or incomplete datasets. In fields like healthcare, biased decisions can have life-altering consequences.
- Approach:
  - Conduct real-world scenario testing to identify and mitigate biases: Simulate various environments and edge cases to understand potential biases.
  - Build fairness metrics into the decision-making framework: Define and measure fairness standards that the AI must adhere to.
  - Establish governance mechanisms to monitor bias in live environments: Create oversight systems that continuously evaluate the AI's fairness.
- Use Case: For autonomous vehicles, ensuring unbiased decision-making is critical in diverse urban environments. For instance, the AI must treat pedestrians equally, regardless of age, ethnicity, or clothing. Rigorous testing in simulated and real-world scenarios, such as heavy traffic or adverse weather, helps identify biases in object detection or decision-making, ensuring the vehicle operates safely and fairly for all road users.

c. Accountability

Generative AI:

- Focus: Accountability lies with developers and organizations for unintended misuse or harmful outputs. Gen AI often lacks control over how users leverage its outputs.

- Approach:
    - Establish clear usage policies and guidelines for end-users: Define acceptable use cases and prohibited applications.
    - Provide tools for users to report harmful or inappropriate outputs: Encourage feedback to identify and address problematic behavior.
    - Maintain a chain of accountability across developers, trainers, and operators: Assign responsibilities across all stakeholders involved in the AI lifecycle.
- Use Case: A content creation platform powered by Gen AI might inadvertently produce offensive or misleading advertisements. By implementing a robust review process for AI-generated outputs and offering clear reporting mechanisms for end-users, the company can address issues promptly. This ensures the platform's outputs align with ethical standards and brand values, safeguarding its reputation.

Agentic AI:

- Focus: Accountability involves the outcomes of autonomous actions, particularly in high-stakes scenarios. Failures can lead to significant operational, financial, or physical harm.
- Approach:
    - Define responsibility for errors or failures, whether human, machine, or both: Clarify who is liable in the event of a system failure.
    - Implement override mechanisms to enable human intervention during critical decision points: Ensure humans can intervene when the AI acts outside expected boundaries.
    - Enforce strict regulatory compliance and liability agreements: Work closely with regulators to define and adhere to accountability standards.
- Use Case: In financial trading, an autonomous trading bot might execute high-risk trades that result in significant losses. By assigning clear accountability to system operators and incorporating real-time monitoring tools, financial institutions can prevent such incidents. Additionally, override mechanisms allow human traders to halt the bot's operations in case of erratic behavior, mitigating potential risks.

d. Risk Management

Generative AI:

- Challenge: Risks include hallucinations, misinformation, and misuse for malicious purposes (e.g., deepfakes). The open-ended nature of Gen AI makes it susceptible to unintended applications.
- Approach:
  - Conduct regular audits to identify and minimize risks: Continuously evaluate the AI's outputs for potential risks.
  - Build safety nets to detect and prevent harmful content generation: Integrate safeguards that flag inappropriate outputs before they reach users.
  - Implement user-facing controls to limit model capabilities in sensitive applications: Allow users to adjust the model's settings to suit their specific needs.
- Use Case: A Gen AI tool used to generate news articles must be configured to avoid producing fake news. For instance, when summarizing political events, the tool should cross-reference credible sources and flag unverifiable claims. This ensures that the generated content maintains journalistic integrity and avoids spreading misinformation.

Agentic AI:

- Challenge: Risks involve physical or financial harm caused by incorrect autonomous actions. These risks can be catastrophic in safety-critical industries like aviation or healthcare.
- Approach:
  - Conduct extensive simulations and real-world testing to identify edge cases: Test the AI under various conditions to understand potential failures.
  - Develop fail-safe mechanisms to ensure safety during failures: Incorporate redundant systems to prevent catastrophic outcomes.
  - Establish robust monitoring systems to detect anomalies in real time: Use advanced tools to identify and respond to unexpected behavior.
- Use Case: For autonomous drones used in logistics, risk management includes planning for hardware failures, such as engine malfunctions or communication loss. Fail-safe mechanisms like parachute systems and alternate

communication channels ensure that the drones can safely land without causing damage to property or people.

e. Ethical and Societal Implications

Generative AI:

- Challenge: Ethical concerns for Generative AI revolve around intellectual property (IP) violations, data privacy, and the creation of harmful or offensive content. Generative AI's ability to produce new and original content presents unique challenges, as the lines between original and derivative works are often blurred.
- Approach:
  - Ensure compliance with IP and copyright laws: Organizations must implement measures to prevent Generative AI systems from infringing on existing intellectual property rights. For example, training data should exclude copyrighted material unless proper permissions are obtained.
  - Use anonymized data during training: Anonymizing data ensures that personal information is not exposed during the model's operation, addressing privacy concerns.
  - Engage ethicists and diverse communities: Collaborating with ethicists and communities ensures that cultural and social sensitivities are respected. This is particularly important when the AI generates content for global audiences.
- Use Case: AI-generated art platforms like DALL-E must ensure that the outputs do not replicate copyrighted works. For instance, if a user prompts the system to create an image in the style of a famous artist, the AI should produce an original interpretation rather than duplicating existing artworks. Regular audits and clear disclaimers to users help navigate these ethical challenges.

Agentic AI:

- Challenge: Ethical concerns for Agentic AI center on safety, fairness, and the moral implications of delegating decisions to autonomous systems. These systems, often operating in critical environments, must align with societal values to avoid harm or misuse.
- Approach:
  - Implement ethical decision-making frameworks: Organizations must define and program ethical boundaries for autonomous systems. For

example, autonomous vehicles should prioritize human safety over property damage in unavoidable accident scenarios.

- Collaborate with regulators and policymakers: Building frameworks for compliance with legal standards ensures alignment with societal expectations and reduces risks of misuse.
- Ensure transparency in trade-offs: Clearly communicate the ethical trade-offs made during system design. For instance, a self-driving car's decision-making algorithm should be transparent to regulators and users alike.

- Use Case: In the context of autonomous drones used in military operations, strict adherence to international humanitarian laws is critical. These systems must be programmed to identify and avoid civilian targets while maintaining accountability mechanisms that allow human operators to override decisions in real time.

f. Scalability of Oversight Implications

Generative AI:

- Challenge: The widespread adoption of Generative AI across industries introduces significant oversight challenges. Due to the vast number of users and use cases, ensuring ethical and appropriate use becomes increasingly difficult, especially for publicly accessible models.
- Approach: Organizations must implement multi-layered oversight systems, including:
    - Proactive Monitoring: Automate monitoring using AI-driven tools to flag suspicious patterns of use.
    - User Authentication: Restrict access to advanced features for verified users.
    - Post-Use Audits: Regularly review logs and outputs to identify trends in misuse.
    - Collaboration with Stakeholders: Work with governments and institutions to develop policies for responsible usage.
- Use Case: In education, an AI-powered content generation tool could be used by students to create essays or assignments. Without proper oversight, the system might be exploited for plagiarism, compromising academic integrity. Institutions can mitigate this by integrating plagiarism detection software and

enforcing clear usage policies, ensuring students use the tool for learning rather than academic dishonesty.

Agentic AI:

- Challenge: Scaling oversight for Agentic AI systems deployed in real-world scenarios, such as fleets of autonomous vehicles or drones, requires robust infrastructure for real-time monitoring and decision validation. The stakes are higher due to the potential physical or financial harm these systems can cause.
- Approach: Oversight mechanisms for Agentic AI require:
  - Centralized Control Systems: Implement command centers that monitor all systems in real time, providing alerts for anomalies.
  - Redundancy Protocols: Ensure fail-safe mechanisms to handle system failures or unexpected behavior.
  - Geofencing and Regulation Compliance: Use geofencing to enforce operational boundaries and align with local regulations.
  - Stakeholder Communication: Establish clear channels for reporting issues and coordinating resolutions among users, regulators, and developers.
- Use Case: In logistics, a company deploying autonomous delivery drones in multiple cities could face regulatory and operational challenges. For example, if a drone encounters restricted airspace or loses connectivity, centralized monitoring systems can reroute it or trigger an emergency landing protocol. Collaborating with aviation authorities and adopting compliance frameworks ensures scalability without compromising safety or trust.

Summary Table of Differences and Approach for Responsible AI

Dimension	Generative AI	Agentic AI
Transparency	Requires explanation of outputs and training data limitations.	Needs interpretability of decision-making processes in real-time.
Bias Mitigation	Focuses on diverse datasets and post-processing to reduce content bias.	Involves fairness metrics and scenario testing for autonomous actions.

Accountability	Primarily involves developers and organizations managing misuse.	Emphasizes liability for autonomous decisions and robust override mechanisms.
Risk Management	Mitigates risks of hallucinations, misinformation, and malicious misuse.	Focuses on fail-safe mechanisms to prevent harm in critical applications.
Ethical Concerns	Addresses IP, copyright, and societal impact of generated content.	Involves safety, fairness, and moral implications of delegating autonomy.
Scalability of Oversight	Monitors diverse use cases and user behaviors across large-scale deployments.	Requires real-time monitoring and fail-safe protocols for high-stakes environments.
Use Cases	Creativc industries, content generation, educational tools.	Autonomous vehicles, healthcare diagnostics, financial trading bots.
Solution Approach	Employs user-facing disclaimers, post-processing tools, and diverse training data.	Utilizes centralized control systems, redundancy protocols, and regulatory compliance.

3. Conclusion

Generative AI and Agentic AI present unique opportunities and risks, requiring tailored Responsible AI approaches. While Gen AI emphasizes content accuracy, bias mitigation, and user oversight, Agentic AI focuses more on safety, autonomy boundaries, and ethical decision-making. Organizations must adopt nuanced frameworks for each technology to maximize their potential while ensuring ethical, safe, and responsible usage.

Chapter 20

Technology and Tools Supporting Responsible AI

Introduction

The journey toward Responsible AI is supported by a growing ecosystem of technology platforms, open-source tools, and emerging innovations. These tools empower organizations to design, develop, and deploy AI systems that align with ethical principles, regulatory requirements, and societal values. This chapter provides an overview of the leading platforms, explores open-source tools for AI fairness and auditing, discusses emerging technologies enhancing AI responsibility, and includes a summary table mapping Responsible AI dimensions to available tools.

Overview of Platforms Supporting Responsible AI

1. AWS (Amazon Web Services)
   - **Amazon SageMaker Clarify**: Detects biases in datasets and machine learning models, offering detailed fairness and explainability reports to support transparency and equitable outcomes.
   - **AWS Data Exchange**: Provides diverse datasets critical for building inclusive AI systems, reducing bias and ensuring broad representation.
   - **AWS Identity and Access Management (IAM)**: Strengthens privacy and security by enabling fine-grained access controls, safeguarding sensitive data.
   - **How It Supports Responsible AI**: AWS tools address fairness, transparency, and privacy through bias detection, data inclusivity, and secure access controls.

2. Microsoft Azure
    - **Azure Machine Learning (AML)**: Offers tools for bias detection, interpretability, and secure model deployment, including a comprehensive fairness dashboard.
    - **Azure OpenAI Service**: Embeds safety mechanisms and compliance monitoring into the use of large language models, promoting ethical AI usage.
    - **Azure Policy**: Enables organizations to enforce Responsible AI standards by defining and monitoring compliance rules across projects.
    - **How It Supports Responsible AI**: Azure's tools integrate fairness, transparency, and accountability, providing a holistic framework for ethical AI development.
3. Google Cloud
    - **What-If Tool**: Visualizes model behavior and evaluates fairness by simulating changes in input data, identifying disparities across demographic groups.
    - **Vertex AI**: Supports the AI lifecycle with fairness testing, explainability features, and robust model deployment capabilities.
    - **Differential Privacy Library**: Ensures user data privacy by introducing noise into datasets, maintaining aggregate insights while protecting individual records.
    - **How It Supports Responsible AI**: Google Cloud tools enhance fairness, transparency, and privacy by combining robust fairness evaluations with privacy-preserving techniques.
4. OpenAI
    - **GPT Models**: Designed with safeguards to minimize harmful outputs and align model behavior with ethical principles.
    - **OpenAI API**: Includes moderation tools to filter content and ensure compliance with Responsible AI guidelines.
    - **Alignment Research**: Focuses on aligning AI systems with human values, promoting ethical and accountable usage.
    - **How It Supports Responsible AI**: OpenAI tools emphasize transparency and accountability, ensuring safe and ethical AI deployment.

Open-Source Tools for AI Fairness and Auditing

1. IBM AI Fairness 360 (AIF360)
   - Detects and mitigates bias in datasets and AI models using a suite of fairness metrics and algorithms. The tool integrates seamlessly into workflows, enabling continuous fairness evaluations.
   - **How It Supports Responsible AI**: Addresses fairness and accountability by providing actionable insights and solutions to mitigate bias.
2. Microsoft Fairlearn
   - Evaluates machine learning models for fairness, offering mitigation algorithms and interactive visualizations to address disparities in model performance.
   - **How It Supports Responsible AI**: Enhances fairness and transparency by enabling organizations to identify and correct biases effectively.
3. Google What-If Tool
   - Analyzes model behavior dynamically, allowing users to test scenarios and evaluate fairness in real time. It's highly accessible through TensorFlow and Jupyter Notebooks.
   - **How It Supports Responsible AI**: Focuses on fairness and transparency by providing tools to evaluate and explain model outcomes.
4. LIME (Local Interpretable Model-Agnostic Explanations)
   - Generates interpretable explanations for AI predictions, fostering stakeholder trust by clarifying decision-making processes.
   - **How It Supports Responsible AI**: Promotes transparency and accountability by making complex model outputs understandable.
5. SHAP (SHapley Additive exPlanations)
   - Uses game theory to provide consistent and interpretable explanations for AI predictions, supporting both local and global explainability.
   - **How It Supports Responsible AI**: Enhances transparency and trust by breaking down decision-making processes into comprehensible components.
6. AI Explainability 360
   - Offers tailored explanation algorithms and tools to meet diverse stakeholder needs, supporting textual, visual, and interactive formats.
   - **How It Supports Responsible AI**: Addresses transparency by enabling organizations to provide clear explanations across various use cases.

Emerging Technologies Enhancing AI Responsibility

1. Federated Learning
    - Decentralizes AI training by keeping data localized, reducing privacy risks while ensuring robust model performance across devices.
    - **How It Supports Responsible AI**: Enhances privacy and accountability by minimizing data transfers and ensuring secure local processing.
2. Differential Privacy
    - Introduces noise into datasets to anonymize individual data points while maintaining aggregate data utility, ensuring compliance with privacy regulations.
    - **How It Supports Responsible AI**: Strengthens privacy and regulatory compliance by providing quantifiable privacy guarantees.
3. Model Agnostic Meta-Learning (MAML)
    - Trains models to generalize across tasks, reducing reliance on biased datasets and improving inclusivity in AI systems.
    - **How It Supports Responsible AI**: Promotes fairness and inclusivity by enabling models to adapt equitably across diverse contexts.
4. AI Ethics-as-a-Service Platforms
    - Offers modular services for bias detection, fairness audits, and ethical compliance, simplifying integration into workflows.
    - **How It Supports Responsible AI**: Streamlines fairness, transparency, and accountability processes by providing ready-to-use ethical AI tools.
5. Explainable Reinforcement Learning (XRL)
    - Enhances the transparency of reinforcement learning systems, making their decision-making processes more interpretable.
    - **How It Supports Responsible AI**: Strengthens transparency and accountability, particularly in autonomous systems.

Summary Table: Responsible AI Dimensions and Tools

Responsible AI Dimension	Supporting Tools and Technologies	How Tools Address the Dimension
Fairness	IBM AIF360, Microsoft Fairlearn, Google What-If Tool	Detect and mitigate biases, evaluate model outcomes across groups.
Transparency	SHAP, LIME, AI Explainability 360, Google What-If Tool	Provide clear explanations of AI predictions and decision-making.
Accountability	OpenAI API, Azure Policy, AI Ethics-as-a-Service Platforms	Embed governance rules, monitor compliance, and enforce ethical usage.
Privacy	AWS IAM, Google Differential Privacy Library, Federated Learning	Safeguard sensitive data and ensure compliance with privacy laws.
Inclusivity	MAML, AWS Data Exchange, Microsoft Fairlearn	Ensure diverse representation in data and model adaptability.
Sustainability	Emerging energy-efficient AI technologies, AI-powered optimization tools	Reduce environmental impact of AI systems.

Conclusion

Technology and tools play a pivotal role in advancing Responsible AI by providing the infrastructure, algorithms, and frameworks needed to align AI systems with ethical and societal values. Platforms like AWS, Microsoft, Google, and OpenAI offer robust solutions for Responsible AI practices, while open-source tools such as AIF360 and SHAP democratize access to fairness and explainability capabilities. Emerging technologies like federated learning and explainable reinforcement learning promise to further enhance AI responsibility. The summary table highlights how these tools align with core dimensions of Responsible AI, enabling organizations to build systems that are ethical, trustworthy, and impactful.

Chapter 21

Best Practices: Implementing Responsible AI

This chapter provides practical guidelines and actionable strategies for organizations to implement Responsible AI. By adopting these best practices, businesses can ensure their AI systems align with ethical principles, regulatory requirements, and societal values while delivering value to stakeholders.

1. Establish a Strong Governance Framework
   - Define Governance Structures
     Establish dedicated governance structures, such as AI ethics boards or committees, to oversee AI projects. These entities should include representatives from diverse backgrounds to provide varied perspectives. Define clear roles and responsibilities for ensuring AI systems are ethical, compliant, and aligned with organizational goals. Document these governance structures and communicate their purpose to all stakeholders.
   - Develop Ethical Guidelines
     Organizations should create a set of ethical principles that guide the development and deployment of AI systems. These principles might include fairness, accountability, inclusivity, and transparency. Ethical guidelines should be practical and actionable, providing teams with clear expectations and scenarios for implementation. Periodically review and update these guidelines to reflect advancements in technology and evolving societal values.

- Ensure Stakeholder Engagement
  Effective governance involves engaging a wide range of stakeholders, including employees, customers, regulators, and community representatives. Regular consultations with stakeholders can provide valuable insights into potential risks and benefits of AI systems. This collaborative approach ensures that AI initiatives address diverse needs and concerns while fostering trust and inclusivity.
- Periodic Audits
  Regular audits help ensure compliance with ethical and regulatory standards. Audits should assess not only technical aspects like model performance but also ethical considerations such as fairness and accountability. Use external, independent auditors where possible to enhance credibility and objectivity. Document audit findings and share them with relevant stakeholders to maintain transparency.

2. Promote Transparency and Explainability
   - Document AI Workflows
     Maintain detailed records of the entire AI lifecycle, including data collection, preprocessing, model development, and deployment. Documentation should highlight key decision points and assumptions made during development. This level of transparency facilitates traceability and simplifies debugging or improvements when issues arise.
   - Enhance Model Explainability
     Explainability tools such as LIME and SHAP can be used to clarify how AI models reach their decisions. These tools can help identify factors influencing outcomes, enabling developers to address biases or errors. Explainability is particularly important in high-stakes applications such as healthcare or finance, where decisions must be easily understood and justified.
   - Communicate Decisions Clearly
     Develop intuitive dashboards, reports, or visualizations to convey AI-driven outcomes to non-technical stakeholders. Tailor communication to the audience, ensuring technical details are accessible to those without specialized knowledge. Providing clear explanations fosters trust and enables stakeholders to make informed decisions based on AI outputs.
   - Foster Trust
     Be upfront about the limitations of AI systems, including potential inaccuracies or biases. Organizations should also communicate any risks

associated with AI deployment. Transparent communication about strengths and weaknesses builds user confidence and mitigates unrealistic expectations.

3. Ensure Fairness and Mitigate Bias
    - Diverse Data Collection
      Ensure datasets are representative of the populations the AI system will serve. This requires proactive efforts to include underrepresented groups or scenarios. Regularly assess datasets for imbalances or gaps, and supplement them as needed to ensure inclusivity. Diverse data reduces the risk of biased outcomes and improves the overall robustness of AI models.
    - Bias Detection and Mitigation
      Use tools like AI Fairness 360 and Fairlearn to systematically identify biases in datasets and models. Bias detection should occur at multiple stages, from data preprocessing to post-deployment monitoring. Once biases are identified, implement corrective measures, such as rebalancing datasets or introducing fairness constraints in model training.
    - Incorporate Fairness Metrics
      Evaluate models using fairness metrics like demographic parity, equalized odds, or disparate impact. These metrics quantify bias and provide measurable goals for improvement. Include fairness metrics in regular performance reports to ensure they are treated as priorities alongside accuracy and efficiency.
    - Continuous Monitoring
      Fairness is not a one-time achievement but requires ongoing vigilance. Monitor deployed models for changes in behavior that could indicate emerging biases. Retrain models as necessary to adapt to new data or contexts, ensuring fairness remains consistent over time.
4. Prioritize Data Privacy and Security
    - Implement Privacy by Design
      Integrate privacy considerations into the AI system from the earliest stages of development. Techniques like data minimization, anonymization, and differential privacy should be standard practices. This proactive approach ensures compliance with privacy laws and protects user trust.

- Secure Data Storage
  Use advanced encryption techniques to protect sensitive data, both at rest and in transit. Implement strict access controls to ensure only authorized personnel can access data. Regularly update security protocols to address evolving threats and vulnerabilities.
- Regular Security Audits
  Conduct comprehensive security audits to identify weaknesses in data handling and system architecture. Audits should cover aspects like data encryption, access control, and vulnerability management. Address findings promptly to prevent data breaches or unauthorized access.
- Compliance with Regulations
  Stay updated on data protection laws such as GDPR, CCPA, or HIPAA, and ensure AI systems comply with these regulations. Assign dedicated personnel to oversee compliance efforts and provide training to teams on regulatory requirements.

5. Foster Accountability and Responsibility
   - Human-in-the-Loop Mechanisms
     Incorporate human oversight into AI systems, particularly for critical decisions like medical diagnoses or loan approvals. Human-in-the-loop systems allow for error correction, ethical judgment, and accountability. Clearly define when and how human intervention is required to prevent over-reliance on automation.
   - Clear Accountability Frameworks
     Assign accountability for each stage of the AI lifecycle, from data collection to deployment. Define who is responsible for addressing issues or failures. This clarity ensures that ethical lapses or technical errors are promptly addressed, reducing the risk of harm.
   - Incident Response Plans
     Develop detailed protocols for handling incidents such as ethical breaches, model failures, or security vulnerabilities. Include steps for identifying the issue, notifying stakeholders, and implementing corrective actions. Regularly test and update these plans to ensure they remain effective.
   - Ethics Training
     Provide ongoing ethics training for employees involved in AI projects. Training should cover topics like bias mitigation, privacy considerations,

and regulatory compliance. A well-informed workforce is essential for embedding ethical practices into AI development.

6. Leverage AI Tools and Technologies
   - Adopt Responsible AI Tools
     Leverage tools like Microsoft's Fairness Dashboard, Google's What-If Tool, or AWS SageMaker Clarify to enhance AI fairness, explainability, and compliance. These tools provide automated solutions for identifying and addressing ethical challenges in AI systems.
   - Use Open-Source Libraries
     Utilize open-source libraries like Fairlearn, AI Fairness 360, or TensorFlow Privacy to incorporate ethical practices into AI workflows. These libraries provide pre-built functionalities for detecting bias, ensuring privacy, and improving transparency.
   - Automate Monitoring and Auditing
     Implement automated systems for continuous monitoring and auditing of AI performance. These systems can detect anomalies, measure compliance with ethical standards, and provide alerts for potential issues. Automation reduces the manual workload while improving oversight.
7. Monitor and Iterate
   - Performance Monitoring
     Regularly track AI system performance using metrics such as accuracy, fairness, and reliability. Monitoring should include technical aspects like system latency and user-centric metrics like satisfaction or trust.
   - Feedback Loops
     Establish mechanisms for collecting feedback from users and stakeholders. Feedback should be analyzed to identify areas for improvement, such as usability, accuracy, or fairness. Actively incorporating feedback ensures the system evolves to meet user needs.
   - Periodic Model Retraining
     Retrain models using updated datasets to ensure they remain relevant and effective. This is particularly important in dynamic environments where data patterns change over time. Retraining also helps mitigate issues like model drift or emerging biases.
   - Learning from Incidents
     Document and analyze incidents such as ethical breaches or technical failures. Use these lessons to refine processes, improve training, and

prevent similar issues in the future. Share learnings across teams to foster a culture of continuous improvement.

8. Align AI with Corporate Social Responsibility (CSR)
    - AI for Social Good
      Explore opportunities to apply AI to societal challenges such as healthcare, education, or environmental sustainability. Initiatives like predictive healthcare models or energy-efficient AI systems demonstrate responsible innovation.
    - Community Engagement
      Engage with local communities to understand their needs and concerns regarding AI systems. Collaborate with community organizations to ensure AI projects align with societal goals and values. Involving communities fosters trust and inclusivity.
    - Environmental Impact
      Design AI systems with sustainability in mind, optimizing for energy efficiency and reducing carbon footprints. Use cloud services with renewable energy commitments and adopt practices like model optimization to minimize resource usage.
9. Industry-Specific Considerations
    - Healthcare
      Ensure AI systems comply with medical regulations and prioritize patient safety and data privacy. Use explainability techniques to help healthcare providers understand AI-driven insights and make informed decisions.
    - Finance
      Focus on transparency, fairness, and compliance with financial regulations. Implement rigorous auditing and bias detection processes to maintain trust and avoid risks like fraud or discrimination.
    - Retail
      Use AI to enhance customer experiences through personalized recommendations or dynamic pricing. However, ensure these practices respect privacy and do not exploit vulnerable groups. Implement transparency measures to build customer trust.

10. Measure and Report Impact
    - Define Key Metrics
      Develop comprehensive metrics to evaluate the impact of AI systems. Metrics should cover ethical dimensions like fairness and inclusivity, as well as business outcomes like efficiency and customer satisfaction.
    - Regular Reporting
      Share progress on AI initiatives through transparent reports or dashboards. Reports should highlight successes, challenges, and areas for improvement. Transparency fosters trust among stakeholders and demonstrates accountability.
    - Third-Party Assessments
      Engage independent auditors or evaluators to validate AI practices. Third-party assessments provide an unbiased perspective and enhance credibility. Use their findings to refine practices and strengthen AI governance.

By following these best practices, organizations can build and maintain AI systems that are ethical, transparent, and aligned with societal values. These guidelines not only mitigate risks but also foster trust, innovation, and long-term success in AI initiatives.

Chapter 22

Governance for Responsible AI

Introduction

Governance is a critical element in ensuring the success of Responsible AI (RAI) initiatives. By establishing robust governance structures, enterprises can ensure that AI systems align with ethical principles, regulatory requirements, and organizational goals. This chapter explores the key stakeholders involved in RAI governance, the principles and frameworks that underpin effective governance, and the integration of RAI governance with other enterprise governance forums.

Stakeholders in Responsible AI Governance

1. Chief AI Officer or Responsible AI Lead
    - **Role**: Oversees the organization's AI governance strategies, ensuring alignment with business objectives, ethical principles, and regulatory requirements. They provide strategic direction for the development and deployment of AI systems.
    - **Why Included**: This role is pivotal in driving Responsible AI initiatives at the executive level, ensuring that governance frameworks are prioritized and that resources are allocated effectively to meet governance objectives.

2. Data Scientists and Machine Learning Engineers
    - **Role**: Responsible for developing, testing, and validating AI models. They ensure that models adhere to ethical and technical standards by embedding fairness, transparency, and accuracy in their design.

   - **Why Included**: These technical experts are essential for translating governance principles into actionable practices. Their involvement ensures that the systems are robust, interpretable, and aligned with Responsible AI principles.
3. Legal and Compliance Teams
   - **Role**: Monitor and enforce adherence to regulations such as GDPR, HIPAA, or industry-specific guidelines. They also assess legal risks and provide guidance on ethical and lawful AI usage.
   - **Why Included**: Their expertise is crucial for navigating the complex regulatory landscape surrounding AI. Including these teams helps organizations avoid potential legal pitfalls and ensures compliance with evolving global standards.
4. Ethics Officers
   - **Role**: Act as advocates for embedding ethical considerations into AI systems. They assess potential societal impacts and ensure that AI use aligns with the organization's ethical values.
   - **Why Included**: Ethics officers bring a critical perspective on fairness, bias, and societal implications. Their role helps organizations navigate moral complexities and build AI systems that are aligned with human-centered values.
5. IT and Security Teams
   - **Role**: Focus on building secure infrastructures for AI systems, implementing data protection protocols, and addressing potential cyber threats.
   - **Why Included**: These teams ensure the resilience and security of AI systems, safeguarding sensitive data and maintaining user trust. Their involvement mitigates risks related to data breaches and operational failures.
6. Business Unit Leaders
   - **Role**: Define the scope of AI applications, ensuring that AI initiatives align with organizational goals and deliver tangible business value.
   - **Why Included**: Their strategic input ensures that AI systems are developed with a clear purpose and that governance practices do not hinder business innovation.
7. External Advisors or Ethics Boards
   - **Role**: Provide unbiased, third-party perspectives on governance practices, ethical considerations, and societal impacts.

- **Why Included**: Their independent evaluations ensure transparency and accountability, offering insights that internal stakeholders might overlook. This broadens the ethical and regulatory scope of governance.

Governance Approach for Responsible AI

1. Establish Governance Bodies
    - Organizations should form dedicated governance teams, such as an AI Ethics Committee or a Responsible AI Governance Board. These bodies oversee AI initiatives, ensuring they align with ethical and regulatory standards.
    - Governance bodies should include representatives from various departments, such as data science, legal, ethics, and business units, to provide a multidisciplinary perspective.
    - Clearly define roles and responsibilities within these bodies to streamline decision-making and accountability. Reporting lines should be transparent to maintain oversight.
2. Develop Policies and Standards
    - Policies should comprehensively cover data collection, preprocessing, model development, deployment, and monitoring. These standards should emphasize ethical principles such as fairness, transparency, and accountability.
    - Align organizational policies with international regulations and industry best practices to maintain compliance and ethical integrity.
    - Regularly update policies to reflect technological advancements and evolving societal expectations.
3. Implement Governance Tools
    - Leverage automated tools for tasks such as fairness audits, explainability, and monitoring compliance with regulations. These tools can identify biases and provide actionable insights.
    - Implement dashboards to track governance metrics, enabling real-time monitoring of AI system performance and ethical adherence.
    - Use tools that integrate seamlessly into existing workflows to ensure efficiency and minimize disruptions.
4. Foster a Culture of Responsibility
    - Organizations should invest in regular training sessions to educate employees about Responsible AI principles and governance practices. Training should be tailored to different roles, ensuring relevance.

- Encourage open dialogue about ethical concerns, providing a safe space for employees to report issues or raise questions.
- Recognize and reward efforts that promote ethical AI practices, reinforcing a culture of accountability and integrity.

5. Conduct Regular Audits
   - Schedule periodic reviews of AI systems to assess their compliance with governance policies and ethical standards. These audits should examine data quality, model behavior, and decision outcomes.
   - Use findings from audits to refine governance frameworks and address any identified risks or gaps proactively.
   - Document audit processes and results to ensure transparency and provide a basis for continuous improvement.

Integration with Other Governance Forums

1. Data Governance Forums
   - Collaboration with data governance forums ensures that data used in AI systems meets quality, privacy, and security standards. Shared practices for data stewardship can enhance consistency across initiatives.
   - Data governance forums can provide expertise on managing data biases and implementing robust data validation protocols, which are critical for Responsible AI.
   - This integration helps create a unified approach to managing data-related risks across the organization.
2. IT Governance Committees
   - Aligning with IT governance ensures that AI systems adhere to organizational standards for infrastructure, scalability, and cybersecurity.
   - IT governance committees can provide guidance on integrating AI solutions into existing IT ecosystems, minimizing operational risks.
   - Joint efforts between IT and RAI governance can strengthen incident response strategies and system resilience.
3. Risk Management Boards
   - Collaboration with risk management teams enables a comprehensive approach to identifying and mitigating AI-related risks. This includes reputational, operational, and compliance risks.
   - Shared risk assessment frameworks can streamline the evaluation of AI systems, ensuring consistency and thoroughness.

- Risk management boards can also provide expertise in scenario planning, helping to anticipate potential failures or ethical breaches.

4. Sustainability Councils
    - Partnering with sustainability councils ensures that AI initiatives align with organizational environmental goals. This includes optimizing energy usage and reducing the carbon footprint of AI operations.
    - Shared metrics and reporting practices enable organizations to track the environmental impact of AI systems effectively.
    - Collaboration fosters innovation in sustainable AI practices, such as adopting green data centers or energy-efficient algorithms.
5. Ethics Committees
    - Ethics committees can provide cross-functional expertise, addressing complex ethical dilemmas that arise during AI development and deployment.
    - Shared governance practices with ethics committees ensure coherence in ethical standards across different organizational initiatives.
    - Regular consultations with ethics committees help maintain a strong ethical foundation for AI governance.

Conclusion

Governance for Responsible AI is a cornerstone of ethical and effective AI deployment. By involving diverse stakeholders, adhering to robust governance principles, and integrating with other governance forums, organizations can ensure their AI systems are aligned with societal values and business goals. A structured governance framework not only minimizes risks but also fosters innovation and trust in AI technologies.

Chapter 23

Challenges in Implementing Responsible AI

Introduction

The rise of Artificial Intelligence (AI) has brought remarkable advancements across industries, but it has also raised critical ethical and operational concerns. Implementing Responsible AI (RAI) practices is crucial to ensure that AI systems align with societal values, promote fairness, and prevent harm. However, organizations face several challenges in adopting and integrating RAI frameworks. This chapter explores these challenges and highlights how enterprise tools from AWS, Google, Microsoft Azure, and other providers, along with advancements in large language models (LLMs), are addressing them through innovative solutions and support systems.

Challenges in Implementing Responsible AI

1. Bias and Discrimination

   AI models trained on historical or imbalanced data often perpetuate biases, leading to unfair outcomes.

   - AI systems rely heavily on the quality and diversity of training data, and when datasets reflect societal biases, these biases are amplified in AI outputs. For example, hiring algorithms may favor certain demographics due to historical hiring patterns embedded in the data.
   - Identifying and mitigating bias requires continuous auditing and updating of data, algorithms, and decision-making processes.

- Large language models (LLMs) are evolving to include bias detection modules that analyze outputs for discriminatory patterns and suggest corrections in real-time.

2. Lack of Transparency and Explainability
   Many AI systems operate as black boxes, making it difficult to understand or explain their decision-making processes.
   - AI users often struggle to comprehend why a system made a particular decision, especially in high-stakes domains like finance or healthcare. This lack of clarity can erode trust and hinder adoption.
   - Developing explainable AI requires integrating tools that can break down complex algorithms into interpretable components for stakeholders.
   - LLMs like OpenAI's GPT models are incorporating features that provide context and reasoning behind generated responses, aiding in explainability efforts.
3. Data Privacy and Security Concerns
   AI systems often require large volumes of sensitive data, raising privacy and compliance issues.
   - Protecting personal data is a legal and ethical imperative, particularly under regulations like GDPR and CCPA. Mishandling data can lead to breaches, financial penalties, and reputational damage.
   - Balancing the need for data to train AI systems with privacy considerations is a persistent challenge.
   - LLMs are incorporating privacy-preserving techniques, such as differential privacy, to ensure sensitive data remains secure during training and inference.
4. Operationalizing Ethical Principles
   Translating high-level ethical principles into actionable, measurable practices is a significant hurdle.
   - While organizations often adopt ethical guidelines, integrating these principles into daily workflows and AI development processes remains challenging.
   - Defining metrics for abstract concepts like fairness, inclusivity, and accountability requires domain expertise and collaboration.
   - LLMs are increasingly being used to draft and test ethical policies, simulate scenarios, and evaluate adherence to Responsible AI principles.
5. Dynamic Regulatory Environment
   AI regulations vary across regions and are constantly evolving.

- Organizations must navigate a complex landscape of local and global regulations, such as the EU AI Act and emerging laws in North America and Asia.
- Keeping AI systems compliant requires adaptive strategies and ongoing monitoring of regulatory changes.
- LLMs assist in compliance by analyzing legal texts, summarizing requirements, and offering actionable insights tailored to specific jurisdictions.

6. Technical and Resource Constraints
   Implementing RAI often requires advanced technical expertise, tools, and financial investment.
   - Small and medium enterprises (SMEs) may lack the resources to build dedicated teams or access state-of-the-art tools for implementing Responsible AI.
   - Advanced techniques for bias mitigation, explainability, and robustness often demand specialized skills.
   - Pre-trained LLMs are democratizing access to RAI by providing off-the-shelf solutions for ethical AI implementation, reducing the need for extensive in-house expertise.
7. Resistance to Change
   Adopting RAI often requires cultural and organizational shifts.

- Resistance can stem from a lack of awareness, fear of increased workload, or skepticism about the value of ethical AI practices.
- Building organizational buy-in requires training, leadership commitment, and showcasing the long-term benefits of Responsible AI.
- LLMs are being used in organizational change management by generating training content, facilitating discussions, and helping leaders craft persuasive messages about the importance of RAI.

Tools and Initiatives Addressing Responsible AI Challenges

1. AWS Initiatives
   AWS offers a suite of tools and frameworks to address Responsible AI challenges:
   - Amazon SageMaker Clarify:
     - Identifies and mitigates bias in machine learning models by analyzing datasets and model predictions. The tool provides fairness

metrics, enabling developers to assess disparities across demographic groups and adjust their models.
    - Supports integration with existing workflows to streamline fairness assessments.
    - LLMs integrated with AWS services help analyze large datasets for hidden biases, improving model inclusivity.
- AWS Data Protection Services:
    - Ensures data privacy through encryption, anonymization, and secure storage. Organizations can safeguard sensitive information while complying with regulations like GDPR and HIPAA.
    - Facilitates privacy-preserving AI development by enabling data minimization and secure collaboration.
- Well-Architected Framework for AI/ML:
    - Offers best practices to integrate RAI principles into AI development workflows. This framework provides actionable guidelines for ensuring security, reliability, and ethical design.
    - Includes reference architectures to simplify implementation.

2. Google Initiatives
   Google has been at the forefront of Responsible AI with its tools and research efforts:
   - What-If Tool:
       - Enables users to visualize, analyze, and test machine learning models for fairness. It offers actionable insights to identify bias and optimize models.
       - Facilitates interactive experimentation with different scenarios to evaluate model behavior across demographic groups.
       - Integrates seamlessly with LLM-driven workflows, providing insights into textual and numerical data biases.
   - TensorFlow Privacy:
       - Implements differential privacy techniques to protect user data during model training. This ensures that individual data points cannot be reverse-engineered from the trained model.
       - Supports building privacy-preserving AI systems that align with legal requirements.
   - Google's AI Principles:

    - Guides teams on fairness, safety, and accountability in AI projects. These principles serve as a foundational framework for ethical AI development.

3. Microsoft Azure Initiatives

   Microsoft Azure's Responsible AI initiatives focus on providing practical tools and frameworks:

   - Azure Machine Learning Responsible AI Dashboard:
     - Offers a comprehensive suite of tools for fairness, interpretability, and error analysis. Developers can evaluate and monitor AI systems throughout their lifecycle.
     - Supports integration with LLM-based applications, enhancing model interpretability and transparency.
   - Fairlearn:
     - An open-source toolkit to assess and mitigate fairness issues in machine learning models. It provides metrics and visualization tools to understand model impacts on different groups.
     - Facilitates iterative model improvement by highlighting areas requiring fairness adjustments.
   - Azure Confidential Computing:
     - Protects sensitive data by ensuring secure processing environments. This enables privacy-preserving AI applications that adhere to strict compliance standards.

4. Other Enterprise Initiatives

   Several other organizations and open-source communities are contributing to Responsible AI:

   - IBM AI Fairness 360:
     - A comprehensive open-source toolkit to detect and mitigate bias in AI systems. It supports various fairness metrics and algorithms for diverse use cases.
     - Works alongside LLMs to ensure that textual and decision-making biases are addressed effectively.
   - Hugging Face Transformers:
     - Integrates bias detection tools to ensure responsible use of pre-trained models. The platform emphasizes transparency and accountability in natural language processing applications.
   - OpenAI:

- Conducts research and publishes guidelines to ensure the ethical use of AI systems. It advocates for public collaboration and transparency in AI development.
- LLMs like ChatGPT and GPT-4 incorporate bias reduction strategies and ethical use policies to mitigate risks.

Case Studies: Addressing Responsible AI Challenges

1. Healthcare: Ensuring Fairness in Diagnostic AI
   - **Challenge**: An AI tool for diagnosing skin conditions was found to underperform on darker skin tones due to biased training data.
   - **Solution**: By using Amazon SageMaker Clarify, the organization audited its dataset and retrained the model with more representative data, achieving equitable accuracy across all demographics. LLMs supported the identification of biases in training data through advanced text and image analysis.
2. Finance: Enhancing Transparency in Credit Scoring
   - **Challenge**: Customers faced difficulties understanding the reasons behind their credit scores.
   - **Solution**: A bank implemented Google's What-If Tool to analyze and refine its credit scoring model, enabling more transparent decision-making processes. LLMs provided explanations for credit scoring criteria, improving customer trust.
3. Retail: Protecting Customer Privacy in Personalization Engines
   - **Challenge**: A retail platform's recommendation system raised concerns about data misuse.
   - **Solution**: Using Azure Confidential Computing, the company ensured secure data processing and compliance with privacy regulations while maintaining personalization features. LLMs enhanced privacy-preserving algorithms to protect sensitive customer data.

Conclusion

Implementing Responsible AI is a complex yet critical endeavor that requires addressing challenges like bias, transparency, privacy, and operational hurdles. Enterprise tools and initiatives from AWS, Google, Microsoft Azure, and other providers, alongside advancements in large language models, play a pivotal role in mitigating these challenges. By leveraging these solutions and fostering collaboration across stakeholders, organizations can build AI systems that are ethical, trustworthy, and aligned with societal values.

Chapter 24

Future Directions for Responsible AI

Introduction

As Artificial Intelligence (AI) continues to permeate every facet of society, the demand for Responsible AI practices has never been greater. The ethical, legal, and social challenges surrounding AI require forward-thinking strategies to ensure these technologies are safe, fair, and beneficial to all. This chapter explores the future directions for Responsible AI implementation, focusing on advancing explainability, achieving global standardization, scaling access to RAI tools, and enabling continuous monitoring and adaptation. These initiatives, supported by advancements in Large Language Models (LLMs) and enterprise tools, pave the way for a sustainable and responsible AI ecosystem.

Advancing Explainability

Investing in Research for Interpretable AI Models

The increasing complexity of AI models, especially deep learning architectures, makes their decision-making processes opaque. To address this, research initiatives are exploring innovative techniques to make AI models more interpretable. Methods such as feature attribution, attention mechanisms, and model-agnostic explanations are being developed to shed light on the inner workings of neural networks. Collaboration between academia and industry is vital, with interdisciplinary approaches combining cognitive science, machine learning, and user experience design. Governments and enterprises are also funding these initiatives to accelerate progress, ensuring that explainability becomes a core feature of AI systems.

Developing User-Friendly Tools for Actionable Explanations

Explainability tools are critical for translating complex model logic into actionable insights. Interactive dashboards, visualization platforms, and simulators are being designed to make AI systems comprehensible to users. LLMs, such as OpenAI's GPT models, play a pivotal role by generating human-readable summaries of model behavior, making technical information accessible to non-expert stakeholders. These tools are especially significant in high-stakes industries like healthcare and finance, where end-users require clear explanations to make informed decisions. Future advancements will focus on making these tools more intuitive and contextually relevant, enhancing trust in AI systems.

Global Standardization

Collaborating on International Frameworks

The fragmented regulatory landscape poses significant challenges for organizations operating across multiple jurisdictions. International bodies such as the United Nations, OECD, and ISO are spearheading efforts to create unified frameworks for ethical AI development. These frameworks aim to provide universally accepted definitions for principles like fairness, transparency, and accountability, ensuring consistency and interoperability. Governments, industries, and non-profits are collaborating to address diverse perspectives, creating a balanced approach that aligns global consistency with local adaptability.

Using LLMs to Harmonize Regulatory Guidelines

Large Language Models are proving instrumental in navigating the complexities of multi-jurisdictional compliance. By analyzing regulatory documents from different regions, LLMs can identify overlaps and discrepancies, providing organizations with a harmonized understanding of global guidelines. These models also generate tailored compliance strategies, enabling businesses to adapt quickly to evolving regulations. For small and medium enterprises (SMEs), LLMs democratize access to legal expertise, making compliance with international standards more achievable.

Scaling Access to Responsible AI Tools

Democratizing Access Through Cloud Platforms and Open-Source Initiatives

The high cost and complexity of Responsible AI tools often limit their adoption to large organizations. Cloud platforms like AWS, Microsoft Azure, and Google Cloud are addressing this by offering scalable, pay-as-you-go solutions tailored for smaller enterprises. Open-source projects such as Fairlearn and AI Fairness 360 provide free and customizable resources to assess and mitigate biases, enabling startups and SMEs to integrate Responsible AI practices without significant financial investment. Training programs and community-driven initiatives further empower businesses by building technical expertise and fostering a culture of ethical AI adoption.

Leveraging LLMs for Training and Development

LLMs are transforming training and development programs for Responsible AI. These models can create personalized educational materials, interactive tutorials, and real-time guidance tailored to the unique needs of organizations. By simulating ethical dilemmas, LLMs enable teams to explore practical scenarios, enhancing their understanding of Responsible AI principles. Additionally, these models can adapt training content to specific industries and use cases, ensuring relevance and actionable insights. As LLMs evolve, they will play an increasingly central role in democratizing access to RAI tools and fostering widespread adoption.

Continuous Monitoring and Adaptation

Implementing Real-Time Monitoring Systems

AI systems operate in dynamic environments where data shifts, user behaviors, and external factors can impact performance. Real-time monitoring systems use advanced analytics to detect anomalies and track changes, enabling organizations to address issues proactively. These systems are essential for industries like transportation and finance, where quick responses to anomalies can prevent costly errors and maintain system reliability. Future developments will focus on integrating predictive analytics to anticipate potential problems, further enhancing the robustness of AI systems.

Enhancing Monitoring with LLMs

LLMs excel at processing and interpreting large volumes of unstructured data, making them invaluable for monitoring AI operations. By analyzing logs, reports, and user feedback, LLMs can identify patterns indicative of system malfunctions or biases. Their ability to summarize and prioritize critical insights enables organizations to focus on pressing issues, improving response times and decision-making. Integrating LLMs into monitoring workflows ensures a proactive approach to Responsible AI, allowing systems to adapt dynamically to evolving ethical and operational standards.

Future Innovations and Opportunities

Advancing Explainability Technologies

Research in explainability is expected to evolve beyond current methods by incorporating multimodal approaches that combine textual, visual, and auditory explanations. For example, an AI system diagnosing medical conditions might provide a textual summary, highlight critical features in medical imagery, and include a voiceover explanation for clinicians. These multimodal techniques will make AI outputs more intuitive, bridging the gap between complex algorithms and user comprehension. Additionally, advances in natural language generation within LLMs will enable the creation of customized explanations tailored to diverse user groups, enhancing accessibility and trust.

Global Collaboration for Ethical AI

International alliances between governments, academia, and industry are pivotal for developing universal Responsible AI standards. These collaborations facilitate knowledge sharing, leveraging the expertise of global thought leaders to address ethical challenges comprehensively. Joint initiatives, such as cross-border research projects and international AI ethics conferences, aim to align regulatory approaches and create interoperable frameworks. By promoting the global adoption of ethical AI practices, these collaborations help ensure that AI technologies benefit humanity universally, while addressing region-specific cultural and legal nuances.

LLMs as Ethical Advisors

Future iterations of Large Language Models are poised to function as real-time ethical advisors, offering instant feedback on the ethical implications of AI systems. For instance, an LLM integrated into an AI development platform could analyze design decisions, flagging potential ethical concerns such as bias or privacy violations. These models will provide organizations with actionable recommendations, enabling them to address issues before deployment. Moreover, LLMs will simulate scenarios to predict the societal impact of AI applications, helping businesses align their innovations with societal values and ethical norms. This proactive approach will play a crucial role in fostering Responsible AI development.

Conclusion

The future of Responsible AI implementation lies in advancing explainability, achieving global standardization, scaling access to tools, and enabling continuous monitoring. Supported by innovations in LLMs and enterprise solutions, these initiatives address the ethical, technical, and operational challenges of AI systems. By embracing these future directions, organizations can build AI systems that are not only innovative but also ethical, trustworthy, and aligned with societal values, ensuring a sustainable and equitable AI ecosystem.

Chapter 25

A Call to Action: Shaping the Future of Responsible AI

Introduction

The rise of Artificial Intelligence (AI) presents transformative opportunities across industries, yet it also introduces challenges that require immediate and collective action. Ensuring that AI serves humanity's best interests demands collaboration among governments, organizations, and individuals. This chapter outlines the imperative for shaping the future of Responsible AI through collaborative efforts, building an innovation ecosystem, and articulating a vision for a sustainable and ethical AI future, highlighting examples from AWS and other industry leaders.

Collaborative Efforts Among Governments, Organizations, and Individuals

1. Governments

   Governments play a pivotal role in establishing the regulatory frameworks that guide Responsible AI. By enacting laws and policies, they ensure ethical AI development and usage.

   - **Global Coordination**: Initiatives like the Global Partnership on AI (GPAI) illustrate the potential for nations to collaborate on cross-border AI regulations, ensuring consistency and shared ethical standards. AWS supports these efforts by offering compliance tools such as AWS Artifact to meet global regulatory standards.
   - **Public Investments**: Governments can fund research in AI ethics and inclusivity. For example, the European Union's Horizon Europe program

includes funding for AI projects that emphasize fairness and accountability.

- **Ethical Standards**: Establishing AI ethics councils, such as Singapore's Model AI Governance Framework, provides oversight to ensure AI systems align with societal values. AWS reinforces this by offering transparency tools like Amazon SageMaker Clarify, which helps detect bias and improve model explainability.

2. Organizations

   Enterprises and nonprofits are at the forefront of AI development and deployment. Their commitment to Responsible AI can influence industry-wide practices.

   - **Internal Governance**: Companies like AWS have implemented internal AI governance structures, emphasizing transparency and ethical practices across their cloud services.
   - **Partnerships**: Collaborating with academia and NGOs enables organizations to share knowledge and co-create ethical standards. AWS actively partners with initiatives like the Partnership on AI to develop best practices for Responsible AI.
   - **Transparency**: Organizations like Twitter have published algorithmic impact assessments to evaluate their AI systems' societal effects. AWS contributes by enabling customers to use tools like Amazon SageMaker Model Monitor to ensure transparency in deployed AI models.

3. Individuals

   Every individual has a role to play in shaping Responsible AI, from consumers to AI practitioners.

   - **Awareness and Advocacy**: Consumers can demand greater transparency and accountability from organizations. For instance, users have pushed platforms like Instagram to clarify their algorithmic ranking processes.
   - **Education and Skills Development**: AWS Training and Certification programs offer resources for developers to learn ethical AI practices, empowering them to build responsible solutions.
   - **Participation in Public Dialogue**: Grassroots movements, such as Algorithmic Justice League, enable individuals to influence policy and promote awareness of AI bias and inequities.

Building an Ecosystem for Responsible AI Innovation

1. Fostering Collaboration Across Sectors
   - **Industry-Academia Partnerships**: Universities like MIT are collaborating with corporations such as AWS to co-create AI solutions that integrate rigorous ethical evaluations. These partnerships accelerate innovation while ensuring ethical oversight.
   - **Consortia and Alliances**: Initiatives like the AI Ethics Impact Group bring together multiple stakeholders to drive collective action and share knowledge on ethical challenges in AI.
   - **Open-Source Initiatives**: AWS contributes to open-source initiatives by providing tools like the Deep Graph Library, which promotes collaboration and innovation in AI research.
2. Establishing Ethical AI Standards
   - **Interdisciplinary Approaches**: Engaging experts from various fields ensures comprehensive ethical standards. For instance, the IEEE Global Initiative on Ethics of Autonomous and Intelligent Systems includes technologists, ethicists, and policymakers.
   - **Certification Programs**: AI certification efforts, like those led by AWS's Well-Architected Framework, create benchmarks to ensure AI systems meet ethical and performance standards.
   - **Standardization Bodies**: Organizations like ISO and IEEE develop globally recognized standards for Responsible AI, enabling consistency across industries and regions.
3. Investing in Responsible AI Research and Development
   - **Government and Private Funding**: Programs like the U.S. National AI Initiative allocate resources to prioritize research in areas such as fairness, explainability, and privacy-preserving AI.
   - **Innovation Hubs**: AWS has established innovation centers worldwide to provide organizations with the tools and support needed to develop Responsible AI solutions.
   - **Focus on Emerging Markets**: Startups in regions like Africa are leveraging AI responsibly to address challenges like healthcare access. AWS supports these efforts through cloud credits and training programs to democratize AI access.

Vision for a Sustainable and Ethical AI Future

1. Ethical AI as a Norm, Not an Exception
    - Future AI systems will embed fairness, transparency, and accountability as fundamental design principles. AWS's Ethical AI framework integrates these principles across its services, ensuring that fairness and accountability are not optional but standard.
    - Organizations will adopt ethical AI principles not as a regulatory requirement but as central to their innovation strategies, fostering trust and long-term growth.
2. Universal Access to Responsible AI Tools
    - Democratizing access to AI tools will empower startups and nonprofits to build solutions aligned with Responsible AI. AWS's Free Tier and programs like AWS Activate provide startups with resources to access advanced AI tools.
    - Educational initiatives like AI4ALL aim to equip underrepresented groups with the skills to develop and deploy ethical AI, promoting inclusivity and innovation.
3. AI for Social Good
    - Responsible AI will tackle pressing global challenges. For example, AWS AI tools are used in disaster response scenarios, such as predicting wildfire spread and optimizing emergency response.
    - Collaborative efforts between governments and the private sector, such as Microsoft's AI for Earth program, leverage AI to address climate change and conservation challenges.
4. Continuous Monitoring and Adaptation
    - AI systems will be dynamically monitored and updated to align with evolving ethical standards and societal expectations. AWS's Amazon SageMaker Model Monitor enables organizations to continuously track model performance and detect drifts.
    - Feedback loops involving diverse stakeholders will ensure systems remain accountable and adaptive to new challenges and opportunities.

Conclusion

Shaping the future of Responsible AI requires a collective effort from all sectors of society. By fostering collaboration, building an ecosystem for innovation, and articulating a clear vision, we can ensure that AI systems align with humanity's values and aspirations. Examples from AWS and other industry leaders demonstrate the potential for ethical AI to drive positive societal change. This is not merely an ethical imperative but a strategic opportunity to harness AI's transformative potential for the greater good. The time to act is now—together, we can build a future where AI is responsible, sustainable, and inclusive for all.

Chapter 26

Responsible AI Use Case Implementation Across Various Industries

As artificial intelligence (AI) technologies proliferate, the demand for Responsible AI practices has become a focal point across industries. This chapter delves into practical implementations of Responsible AI in key sectors, demonstrating how ethical, transparent, and accountable AI systems can address unique challenges while driving innovation and societal benefits.

1. Healthcare

1.1 Enhancing Diagnostics

AI systems in healthcare are revolutionizing the diagnostic process by harnessing the power of vast and diverse datasets. These systems ensure accurate results by reducing the potential for human error and minimizing biases. For instance, AI can analyze medical images, such as X-rays and MRIs, to detect anomalies that may go unnoticed during manual reviews. Explainable AI tools provide doctors with insights into the rationale behind diagnostic outputs, fostering trust and enabling informed decision-making. Additionally, these systems prioritize data privacy by employing robust encryption techniques and adhering to global regulations like HIPAA and GDPR, ensuring patient information remains secure. By integrating fairness measures, diagnostic tools address disparities in healthcare access and outcomes across different demographics.

1.2 Personalized Treatment Plans

Personalized treatment plans are a cornerstone of modern healthcare. AI systems analyze a patient's genetic information, lifestyle choices, and environmental factors to create tailored medical solutions. These systems reduce the risk of adverse reactions by identifying specific drugs and interventions that align with a patient's unique profile. Ensuring fairness, AI avoids biases that might exclude underserved populations from receiving optimal care. Transparency is critical in this process; patients and practitioners need to understand how recommendations are derived, which boosts confidence in AI-driven solutions. Moreover, compliance with ethical guidelines ensures the responsible use of sensitive health data while fostering collaboration between AI tools and healthcare providers to improve patient outcomes.

2. Financial Services

2.1 Fraud Detection

AI-powered fraud detection systems are indispensable in combating the growing sophistication of financial crimes. These systems continuously monitor transactions to identify irregularities indicative of fraud, such as unusual spending patterns or anomalous account activities. By using advanced machine learning algorithms, they can adapt to evolving threats in real-time. Explainable AI ensures that alerts are well-documented, enabling financial institutions to act swiftly while maintaining transparency with customers and regulators. Moreover, Responsible AI frameworks prioritize user privacy by implementing stringent security measures, ensuring compliance with laws like GDPR and PCI DSS. This holistic approach not only safeguards assets but also reinforces trust in financial systems.

2.2 Risk Assessment and Credit Scoring

Credit scoring models powered by AI have transformed the lending landscape by evaluating borrower profiles with unparalleled accuracy. These systems analyze a multitude of data points, including financial history and economic trends, to assess creditworthiness. Fairness is paramount in ensuring that historical biases, such as those against specific demographics, do not influence decisions. Transparency allows customers to understand their credit scores and the factors affecting them, fostering trust and accountability. Additionally, rigorous documentation of AI models ensures adherence to financial regulations and facilitates audits. This Responsible AI implementation enables equitable access to credit while driving financial inclusion.

3. Retail and E-Commerce

3.1 Customer Personalization

AI has revolutionized the retail experience by delivering highly personalized customer journeys. By analyzing purchase history, browsing behavior, and preferences, AI systems recommend products tailored to individual tastes. However, Responsible AI ensures these recommendations are inclusive and free from biases, creating equitable opportunities for all consumers. To address privacy concerns, data anonymization techniques are employed, and explicit consent is sought before using personal information. Transparency initiatives, such as providing insights into how recommendations are generated, further enhance customer trust. This balanced approach drives sales and strengthens customer loyalty while adhering to ethical principles.

3.2 Supply Chain Optimization

Supply chain management has benefited immensely from AI-driven insights. By analyzing historical sales data and market trends, AI systems accurately predict demand, preventing overproduction and reducing waste. These systems also optimize logistics by identifying the most efficient transportation routes, minimizing costs and environmental impact. Responsible AI ensures transparency in decision-making, enabling stakeholders to understand and validate AI recommendations. Furthermore, incorporating sustainability measures, such as promoting eco-friendly shipping practices, underscores the commitment to responsible operations. This approach not only enhances efficiency but also aligns with global efforts toward sustainable development.

4. Manufacturing

4.1 Predictive Maintenance

Predictive maintenance powered by AI reduces downtime and operational disruptions in manufacturing. By monitoring equipment performance in real time, AI systems identify signs of wear and tear before failures occur. Explainable AI provides technicians with clear insights into the underlying issues, facilitating timely interventions. These systems also prioritize fairness by ensuring that predictive models are trained on diverse datasets, avoiding biases that could affect specific equipment types. Moreover, adopting Responsible AI practices includes minimizing resource wastage and energy consumption, aligning with sustainability goals. This approach ensures uninterrupted operations while fostering ethical practices in manufacturing.

4.2 Quality Assurance

AI-driven quality control systems enhance product integrity by detecting defects with high precision. These systems analyze production line data to identify inconsistencies, ensuring only high-quality products reach consumers. To maintain accountability, AI models document their decision-making processes, allowing manufacturers to validate compliance with industry standards. Responsible AI also emphasizes reducing material waste during production, promoting eco-friendly practices. By integrating these principles, manufacturers not only improve efficiency but also uphold their commitment to ethical and sustainable operations.

5. Education

5.1 Personalized Learning

AI-powered education platforms adapt learning materials to the unique needs of each student, improving engagement and outcomes. These systems analyze learning patterns to identify strengths and weaknesses, tailoring content and pacing accordingly. Fairness initiatives ensure that all students, regardless of background, receive equitable access to resources. Transparency tools provide educators and students with insights into how AI systems make decisions, fostering trust and collaboration. Additionally, Responsible AI practices prioritize data security, ensuring compliance with regulations and safeguarding student information. This personalized approach transforms education into a more inclusive and effective experience.

5.2 Administrative Automation

Administrative tasks in educational institutions, such as admissions and scheduling, are streamlined using AI. These systems process large volumes of applications efficiently, ensuring fair evaluation criteria. Explainable AI tools enable administrators to understand decision processes, reducing potential biases. Moreover, Responsible AI ensures that sensitive student data is handled securely and transparently. By automating routine tasks, educators can focus more on teaching and mentorship, enhancing overall institutional effectiveness while adhering to ethical standards.

6. Public Sector

6.1 Smart Cities

Smart city initiatives leverage AI to improve urban living conditions through efficient resource management and infrastructure optimization. For example, AI systems analyze traffic patterns to reduce congestion and emissions, enhancing environmental sustainability. These systems also ensure fairness in the allocation of resources, such as housing and public utilities, addressing disparities. Transparency measures, such as public dashboards, keep citizens informed about AI-driven initiatives, fostering trust. By integrating Responsible AI principles, smart cities achieve a balance between technological innovation and ethical governance.

6.2 Disaster Response

AI enhances disaster management by predicting natural calamities and aiding in resource distribution. These systems analyze meteorological and geological data to provide early warnings, saving lives and minimizing damage. Responsible AI ensures equitable allocation of aid, prioritizing the most vulnerable populations. Additionally, maintaining transparency in decision-making processes builds trust with affected communities. This ethical approach to disaster response strengthens resilience and preparedness.

7. Energy and Utilities

7.1 Renewable Energy Optimization

AI systems optimize renewable energy generation by forecasting demand and balancing supply. These systems analyze weather patterns to maximize the efficiency of solar and wind energy production. Responsible AI practices ensure transparency in investment decisions, encouraging stakeholder confidence. Additionally, integrating sustainability measures, such as reducing waste in energy distribution, aligns with global green energy goals. This approach accelerates the transition to cleaner energy solutions while maintaining ethical standards.

7.2 Grid Management

Energy grids benefit from AI-driven real-time monitoring and predictive analysis. These systems detect anomalies and prevent outages, ensuring stable energy delivery. Accountability mechanisms, such as detailed logs of AI decisions, enhance trust among stakeholders. Fairness initiatives ensure equitable energy access, addressing disparities in underserved regions. By adhering to Responsible AI principles, grid management becomes more reliable and inclusive.

8. Transportation

8.1 Autonomous Vehicles

Autonomous vehicles rely on AI for navigation and decision-making, ensuring safe and efficient travel. Responsible AI frameworks prioritize safety by continuously testing and refining algorithms to prevent accidents. Fairness measures address biases in training data, ensuring consistent performance across diverse environments. Transparency tools provide insights into how decisions are made, enhancing public trust in autonomous technologies. By adhering to ethical standards, these vehicles contribute to safer and smarter transportation systems.

8.2 Logistics and Delivery

AI optimizes logistics by determining the most efficient delivery routes, reducing costs and environmental impact. These systems enhance transparency by providing real-time updates to customers on delivery statuses. Fairness practices ensure equitable service availability, avoiding biases in geographic coverage. By integrating Responsible AI, logistics operations achieve higher efficiency while promoting sustainability and ethical practices.

Conclusion

Implementing Responsible AI across industries requires a balance of innovation and adherence to ethical principles. Each sector's unique challenges and opportunities highlight the importance of fairness, transparency, accountability, privacy, and sustainability in AI systems. By adopting these principles, organizations can not only drive progress but also build trust with stakeholders and ensure long-term success in an AI-driven world.

Glossary

Responsible AI Terms

This chapter provides definitions and explanations of key terms and concepts in Responsible AI. Understanding these terms is essential for implementing ethical, transparent, and effective AI systems.

A

Accountability

Accountability refers to the obligation of individuals, organizations, or systems to ensure that AI systems operate ethically and transparently. It involves implementing mechanisms to trace decision-making processes and taking responsibility for any harm caused by AI systems. Accountability also includes establishing clear roles and responsibilities to prevent ethical lapses.

Algorithmic Bias

Algorithmic bias occurs when AI systems produce unfair or discriminatory outcomes due to biased data, design, or implementation. This bias can reinforce existing inequalities and lead to negative social impacts. Addressing algorithmic bias involves understanding its sources and implementing strategies to mitigate it.

Artificial Intelligence (AI)

AI refers to the simulation of human intelligence by machines to perform tasks like learning, reasoning, and decision-making. It encompasses various subfields, including machine learning, computer vision, and natural language processing. AI systems aim to solve complex problems and automate repetitive processes effectively.

B

Bias Mitigation

Bias mitigation involves identifying and reducing biases in AI models and datasets to ensure fair outcomes. Techniques include re-sampling data, re-weighting algorithms, or adopting fairness constraints during training. This process is critical for building equitable AI systems that serve diverse user groups.

Black Box Models

Black box models are AI systems whose inner workings and decision-making processes are not easily understandable by humans. This lack of transparency raises concerns about trust, accountability, and fairness. Efforts to address this include developing interpretable models and implementing explainability techniques.

C

Compliance

Compliance refers to adhering to laws, regulations, and ethical guidelines governing AI development and use. It ensures that AI systems align with standards like GDPR or the AI Act. Compliance safeguards user rights and builds public trust in AI technologies.

Consent

Consent is the process of obtaining explicit permission from individuals to use their data in AI systems. It ensures that users are informed about how their data will be processed and have the option to opt out. Respecting consent is a cornerstone of ethical AI practices.

Critical Decision-Making AI

Critical decision-making AI involves applications where AI systems are used for high-stakes decisions, such as medical diagnoses or loan approvals. These systems require robust design, transparency, and human oversight to minimize risks and ensure ethical outcomes.

D

Data Privacy

Data privacy involves protecting personal and sensitive information from unauthorized access or misuse. AI systems must implement strong encryption, anonymization, and access control measures to maintain user trust and comply with legal requirements.

Data Governance

Data governance is a framework for managing the availability, integrity, security, and usability of data used in AI systems. Effective data governance ensures compliance with regulations and facilitates the creation of high-quality datasets for training AI models.

Diversity in AI

Diversity in AI refers to ensuring that AI systems are designed to represent and serve people from various backgrounds. This includes using diverse datasets and involving multidisciplinary teams. Promoting diversity helps reduce biases and create inclusive technologies.

E

Ethical AI

Ethical AI focuses on developing and deploying AI systems that align with societal values and moral principles. It addresses issues such as fairness, accountability, and transparency to ensure AI systems benefit society while minimizing harm.

Explainability

Explainability is the ability to make AI decisions understandable to humans. It enhances trust by providing insights into how and why an AI system arrived at a specific decision. Techniques like LIME or SHAP are often used to improve explainability.

F

Fairness

Fairness in AI ensures that systems do not produce or reinforce discrimination or inequality. Achieving fairness involves addressing biases in data, algorithms, and outcomes. It is a key ethical principle that guides responsible AI development.

Frameworks for Responsible AI

Frameworks for responsible AI provide structured guidelines and tools to ensure ethical AI practices. Examples include AI Fairness 360 and Fairlearn, which offer resources for bias detection, fairness metrics, and compliance assessment.

G

Governance

Governance in AI refers to the structures and processes that ensure AI systems operate ethically and align with organizational and societal goals. It includes policies for accountability, fairness, and transparency across the AI lifecycle.

Generalization

Generalization is the ability of an AI model to perform well on new, unseen data. It ensures that the model learns underlying patterns rather than memorizing specific details, making it robust for real-world applications.

H

Human-in-the-Loop (HITL)

Human-in-the-loop systems incorporate human oversight into the AI decision-making process. This approach ensures accountability, improves accuracy, and allows for intervention in high-stakes scenarios. HITL is especially important in applications like healthcare and autonomous vehicles.

Hyperparameter Tuning

Hyperparameter tuning is the process of optimizing the settings of an AI model to enhance its performance. It involves adjusting parameters such as learning rate, batch size, or regularization techniques to achieve better results.

I

Interpretability

Interpretability refers to the degree to which humans can understand the workings of an AI model. It is critical for ensuring transparency and trust, especially in applications that affect human lives or involve regulatory requirements.

Inclusive AI

Inclusive AI focuses on designing systems that cater to the needs of diverse populations. It emphasizes accessibility, cultural sensitivity, and fairness to ensure that AI benefits everyone equally.

L

Legal Compliance

Legal compliance ensures that AI systems adhere to relevant laws and regulations, such as GDPR for data protection. It minimizes legal risks and promotes ethical practices in AI development and deployment.

Liability

Liability in AI involves determining who is legally responsible for the outcomes and impacts of AI systems. Clear liability frameworks are essential for addressing disputes and ensuring accountability.

M

Model Governance

Model governance encompasses policies and practices for managing AI models throughout their lifecycle. It includes version control, auditing, and compliance to ensure ethical and reliable AI systems.

Monitoring in AI

Monitoring involves continuously observing AI systems to detect anomalies, biases, or performance issues. Proactive monitoring ensures that AI systems remain effective and compliant over time.

N

Non-Discrimination

Non-discrimination is a principle aimed at ensuring AI systems do not disadvantage any individual or group unfairly. It involves careful design, testing, and validation to prevent biases in AI applications.

Natural Language Processing (NLP)

NLP is a branch of AI focused on enabling machines to understand, interpret, and generate human language. Applications include chatbots, translation services, and sentiment analysis.

P

Privacy by Design

Privacy by design is an approach that incorporates privacy considerations into AI systems from the outset. It emphasizes proactive measures, such as data minimization and encryption, to protect user data.

Provenance of Data

Provenance of data refers to the documentation of the origin, history, and movement of data used in AI systems. Maintaining data provenance ensures transparency and supports regulatory compliance.

R

Responsibility

Responsibility in AI involves ensuring that systems operate in a manner consistent with ethical and societal norms. It includes accountability mechanisms, ethical guidelines, and stakeholder engagement to build trust.

Robustness

Robustness refers to the ability of an AI system to perform reliably under various conditions, including adversarial scenarios or unexpected inputs. Building robust systems is crucial for mission-critical applications.

S

Sustainability in AI

Sustainability in AI focuses on designing systems that are energy-efficient and environmentally friendly. It involves optimizing algorithms and infrastructure to reduce carbon footprints and promote eco-friendly practices.

Security in AI

Security in AI involves protecting systems and data from unauthorized access, tampering, or attacks. Robust security measures, such as encryption and secure access controls, are essential for maintaining trust and compliance.

T

Traceability

Traceability is the ability to track the origin, processing, and decisions of an AI system. It supports accountability and helps address issues such as bias or errors in AI outputs.

Transparency

Transparency involves making AI processes, decisions, and data usage clear and accessible to stakeholders. It is critical for building trust and ensuring ethical practices in AI applications.

U

Unintended Consequences

Unintended consequences are unexpected outcomes resulting from the deployment of AI systems. Addressing these requires proactive risk assessment and iterative development processes.

User-Centric Design

User-centric design prioritizes the needs and preferences of end-users in AI development. It involves engaging stakeholders throughout the design process to ensure the technology is intuitive and beneficial.

V

Verification

Verification is the process of ensuring that an AI system meets its intended specifications and requirements. It involves rigorous testing and validation to confirm the system's functionality and reliability.

Vulnerability in AI

Vulnerability in AI refers to weaknesses in systems that can be exploited or lead to failure. Addressing vulnerabilities requires robust security practices and continuous monitoring.

W

Whistleblowing in AI

Whistleblowing in AI involves reporting unethical or illegal activities in the development or use of AI systems. It is a safeguard to ensure that ethical standards are upheld in AI projects.

Workforce Impact of AI

Workforce impact of AI refers to the changes that AI brings to employment, job roles, and skill requirements. Preparing the workforce for these changes is essential to minimize disruptions and maximize benefits.

X

Explainable AI (XAI)

Explainable AI (XAI) is a subset of AI focused on improving the interpretability and transparency of AI models. It enables stakeholders to understand and trust AI decisions, particularly in high-stakes scenarios.

This glossary serves as a foundational reference for practitioners, policymakers, and academics in understanding and applying Responsible AI principles.

Made in United States
North Haven, CT
04 November 2026

21766924R00100